# Smarter Solar Systems: Optimizing Power Flow and Backup with Advanced Inverters

Nain

# Table of Contents

# 1.  Introduction

## 1.1    Background to the research

In grid-tied residential systems, a solar panel – which provides DC power – can be directly connected to the grid through a DC-AC power converter and provide energy to residential AC loads and reduce the monthly electricity bill. When the grid-goes offline, however, it is required that the entire system goes offline. Solar energy also isn't always available in times such as long winters with little to no sunlight. It therefore is essential to store the energy when it's available and not needed – to be used later when the grid is offline or during winter periods of little sunlight – this is where storage devices are needed. Grid-tied systems then have a storage solution where the solar panel is connected to a storage device and the grid. Batteries are among the storage devices used to store this renewable energy. Since batteries also provide DC power, like PV panels, a power conditioning system (PCS) inverter is used to convert the DC power into AC power. These batteries can be recharged by the grid, through the inverter during night times. Therefore, the inverter requires bidirectional capability to charge the battery bank. Most efficient bidirectional inverters used today are based on two-stage architecture, which involves using a bulky DC-link capacitor. The control strategies used in these inverters are incapable of fully compensating for battery and grid complex disturbance and variations.

## 1.2    Objectives of this research

### 1.2.1    Purpose of the research

The purpose of this research is to analyse and design an efficient $220V$ grid tied, bidirectional inverter and its control strategy for a low voltage, $60V$ battery powered, $1kW$ residential PV system with storage.

### 1.2.2    Problems to be investigated

The problems to investigated include: several battery charging and discharge characteristics to select the most appropriate battery; several PCS architectures to select the most appropriate architecture and front-end converter; the bi-directional PCS's abilities to step up low voltage and reverse power; the dynamics of the bidirectional PCS, to develop a control strategy that allows DC-AC and then AC-DC power flow transition under complex battery or grid disturbance and variations.

## 1.3    Scope and Limitations

This research is focussed on the design and experimental verification of a newly proposed bidirectional inverter topology for battery storage systems. The developed system is limited to a single-phase system and may require further modifications before it can be implemented in a three-phase system.

## 1.4    Plan of development

Chapter 1 is the introduction. Chapter 2 reviews the literature in battery connected systems. Chapter 3 presents a thorough analysis of the inverter being designed, by presenting topology and analysing the modes of operation and power transfer; the effects of switching dead time and parasitic effects; the performance assessment; the efficiency and finally, the dynamics of the inverter. Chapter 4 then presents the design of the inverter by starting with the battery selection, sizing of the components, switch selection and design of the lead compensators. Chapter 5 details the simulated results of both the DC-AC and AC-DC power flows of the bi-directional inverter. Chapter 6 shows and discusses the experimental results. Further discussion is provided in chapter 7 by summarizing the key findings across the simulated and experimental results in relation the theory. Chapter 8 draws conclusions based on the discussions and chapter 9 makes recommendations based on the conclusions that were drawn.

# 2.    Literature Review

This section details the literature review for the proposed topology design. It starts off with a broad overview of residential LV grid-tied systems and their power requirements. It then reviews the various storage options for residential applications. It then reviews the various power conditioning systems (PCS) and compares them for the best architecture. Lastly, control techniques for the PCS were reviewed. All the figures and tables were redrawn from their original sources for coherence.

## 2.1    Residential low voltage grid-tied systems

Most grid-tied residential systems make use of solar and or wind energy to reduce the monthly spending bill by using solar panels to supplement the energy provided by the grid. The cost reduction can be as much as $60 - 80\%$ depending on the size of the residence and type of solar system installed [1]. During times when solar energy is mostly available, the residential loads are supplied by the solar panel and during the night, the grid is used to provide the energy. These systems either have storage or no storage options.

### 2.1.1    *Grid-tied systems with no storage option*

These systems are directly connected to the grid. The DC power provided by the solar panels is connected directly to a power conditioning system (PCS) which is then connected to the grid. This set up as shown in Figure 2-1.

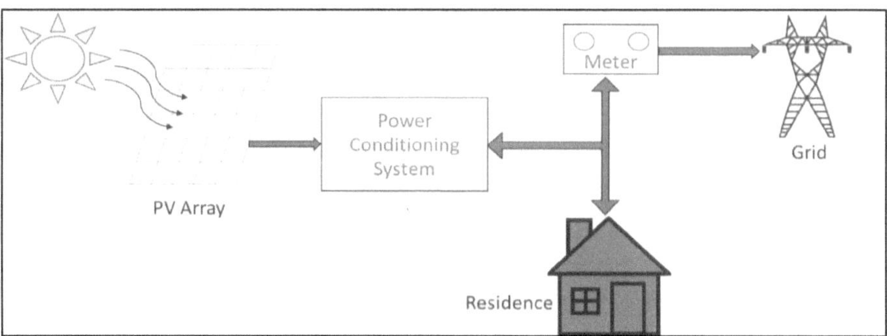


**Figure 2-1 Grid-tied residential system with no storage option**

The solar panel is directly converted into AC power from a PCS. The PCS steps up and converts the DC output voltage into AC voltage to match the grid and residential load standards [2]. When the solar energy is unavailable at night or during winter periods, of little to no sunlight, the grid provides the required energy. This configuration has the advantage of being cheap since no storage option is used. The major issue with this system is that it doesn't reduce the monthly bill by a large factor since, during peak times at night, there is no solar and the grid provides power; and when the grid goes offline, it is a requirement that the entire solar system also be taken offline because if it isn't taken offline, the power generated by the panels can cause shock to any personnel that may be repairing any damage along the grid. During this time, the residence will be without power.

### 2.1.2    Grid-tied systems with storage

To solve the issues of the directly connected grid-tied PV systems discussed in the previous section, storage options are included within the system as shown in Figure 2-2. The PV is connected to the storage device first and then the storage device is connected to the PCS.

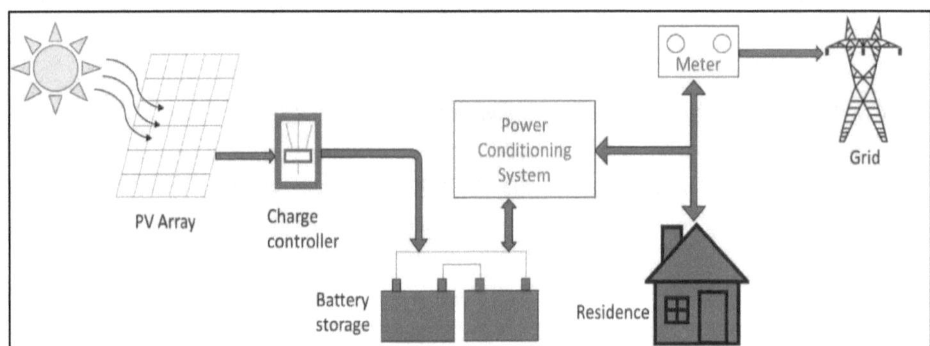


**Figure 2-2 Grid-tied system with storage option**

The solar panels are connected directly to a charge controller. The charge controller can either be MPPT or PWM. These are used to ensure that the storage device is charged efficiently such that, when the PV panels do not generate enough power for the load requirements, the storage device can provide the extra power needed [2]. The battery storage system is connected to a critical loads panels and not the main panel because the power provided by the storage backup is lower than the power provided by the grid. When the grid goes offline, the energy stored in the storage element can be sued to power the critical loads within the resident.

### 2.1.3 Low voltage residential load power and grid quality requirements

The power and its quality required by residential loads in a grid tied system was reviewed.

### i. Power requirements

PV systems with battery backup are only used to power the critical loads within a residence. The critical loads within a residence are typically lights, TV's, cell phone or laptop chargers and fans. These loads need to be powered only for limited time through the battery backup system. Powering 5 incandescent light bulbs, a TV and charging 1 phone will consume about 800W for 3 hours – during load shedding instances, which typically last for approximately 2 and a half hours. A 1kW PV system with battery backup would be enough for this – but larger residences may require about 5kW during these times. A 1kW system with battery backup costs between R70,000 and R90,000 depending on the installation company and solar cells used. A 5kW system costs between R400,000 and R450,000. Not many residents can afford 5kW systems or above, especially since typical residences have basic critical loads – hence, a 1kW system is a popular choice for average residents and will be the focus of the book. For a  PV system with a 1kW max power, 60V storage backup system, the number of PV cells connected in series needed to ensure that the battery pack is fully charged is 216 – which is six 12V PV panels connected in series and will have an open circuit voltage of 108V [3]. When this PV array is connected to a load the open circuit voltage will drop to about 72V – this will be enough to charge a 60V storage backup system. Typical 1kW inverters are rated between 12V to 60V and are usually priced R2,000 and R7,000 making them affordable for most residents interested in renewable cost saving alternatives. Table 2-1 shows the typical power require-ments for an EN 501060 [4] Standard grid-tied PV system with a storage option.

Table 2-1 Typical 1kW PV grid-tied system power requirements

| Identification | PV Panel [3] | Battery [5] | Inverter [6] | Grid [4] |
|---|---|---|---|---|
| **No. of Units** | Six 12$V$ Panels | 1 | 1 | |
| **Rating** | 1200$W$ | 1000$W$ | 1000$W$ | |
| **Voltage** | 72$V$ nominal | 60$V$ | 60$V$ | 220$V$ |
| **Frequency** | | | 50$Hz$ | 50$Hz$ |
| **Power Factor** | | | 0.8 | 0.8 |
| **Manufacturer** | ABC | XYZ | DEF | |

The inverter is the PCS and the battery is the storage element. Specifications for standalone inverters are based on grid-standards because residential appliances were made to fitted to the grid.

*ii.* ***Power quality requirements***

The power quality expected from the inverter to be designed was also reviewed.

Table 2-2 Grid standard steady state voltage characteristics [4]

| Parameter | Supply Voltage Characteristic according to EN 501060 |
|---|---|
| Line frequency | LV, MV: mean value of fundamental measured over 10s <br> $\pm1\%$ $(49.5 - 50.5\ Hz)$ for 99,5% of the week <br> $-6\%$ $(47 - 52\ Hz)$ for 100% of the week |
| Voltage regulation | LV, MV: $\pm10\%$ for 95% of the week, <br> mean 10 minutes RMS values |
| Total Harmonic Distortion | <8% |

Table 2-2 shows the quality of the voltage at the output as per the EN 501060. The inverter to be designed will use these power quality requirements as a standard.

The harmonic standard for residential loads is far more complex and is defined for different classes of loads in the EN 61000-3-2 standard [7] namely, class A, B, C and D.

Table 2-3 Harmonic standard for Class A residential loads [7]

| Harmonic order<br>n | Maximum permissible<br>harmonic current<br>A |
|---|---|
| Odd harmonics | |
| 3 | 2.3 |
| 5 | 1.4 |
| 7 | 0.77 |
| 9 | 0.40 |
| 11 | 0.33 |
| 13 | 0.21 |
| $15 \leq n \leq 39$ | 0.15·8/n |
| Even harmonics | |
| 2 | 1.08 |
| 4 | 0.43 |
| 6 | 0.30 |
| $8 \leq n \leq 40$ | 0.23·8/n |

The quality standard for class A equipment as shown in Table 2-3 defines loads such as, audio equipment and light dimmers. For class B equipment, which includes portable tool and arc welding equipment, the class A standard is multiplied by a 1.5 factor [7].

**Table 2-4 Harmonic standard for Class C residential loads [7]**

| Harmonic order $n$ | Maximum permissible harmonic current expressed as a percentage of the input current at the fundamental frequency % |
|---|---|
| 2 | 2 |
| 3 | $30 \cdot \lambda$ * |
| 5 | 10 |
| 7 | 7 |
| 9 | 5 |
| $11 \leq n \leq 39$ (odd harmonics only) | 3 |
| * $\lambda$ is the circuit power factor | |

Table 2-4 defines the standard for class C loads which include lighting equipment. For class C loads with a real power rating of greater than 25W, Table 2-4 is used. But for class C loads with power ratings of less than 25W, the 3rd harmonic current must be less than 86%; 5th less than 61% of the fundamental current [7].

**Table 2-5 Harmonic standard for Class D residential loads [7]**

| Harmonic order $n$ | Maximum permissible harmonic current per watt mA/W | Maximum permissible harmonic current A |
|---|---|---|
| 3 | 3.4 | 2.30 |
| 5 | 1.9 | 1.14 |
| 7 | 1.0 | 0.77 |
| 9 | 0.5 | 0.40 |
| 11 | 0.35 | 0.33 |
| $13 \leq n \leq 39$ (odd harmonics only) | 3.85/n | See table 1 |

Table 2-5 defines the standard for class D loads which include, personal computers, radios and monitors with input power of less than 600W [7]. Considering a class D residential load such as a drill Figure 2-3 shows the surge power of the load when it is connected to the grid and inverter.

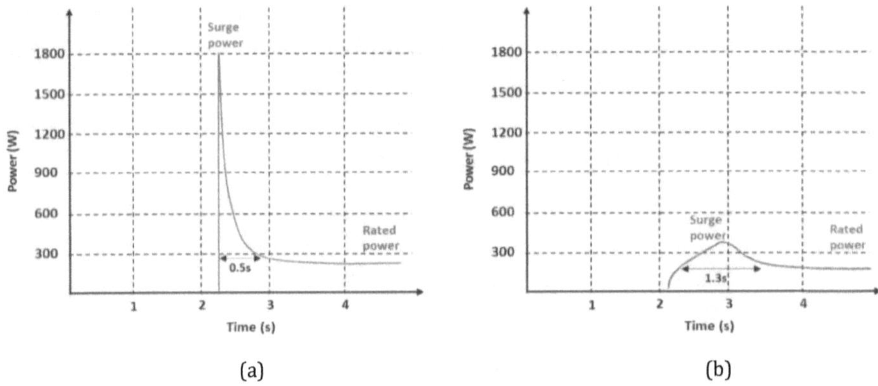


**Figure 2-3 Surge power of residential class D load (a) grid-connected (b) inverter connected**

Figure 2-3 (a) shows that the load connected to the grid has a large surge power that is eight times larger than the power rating of the load. When connected to the inverter, Figure 2-3 (b) shows that the surge power is only 1.42 larger than the rated power of the load. Although this surge power is lower, its settling time is almost 3 times slower than the grid connected load. Therefore, it is advantageous to run household loads on an inverter rather than the grid [8].

## 2.2    Energy storage

Having reviewed that the best kind of system is a grid-tied system with storage option, energy storage options were reviewed. There are several types of renewable storage devices such as super capacitors – which are made of an electric double layer (EDL) [9]. These are only useful for short bursts of power and are therefore, used mostly in start/stop applications that require a lot of charge/discharge cycles during engine operation [9]. Flywheels are another source – they store energy in the form of kinetic energy as they rotate. Like super capacitors, can provide short bursts of power. Batteries are most used storage devices because they can provide a steady stream of power required for residential applications. As such, super capacitors and flywheels are only used to supplement batteries at times when a short burst of power is required [9]. Batteries were then reviewed in terms of their structure, charge and discharge characteristics.

## 2.2.1    *Chemical introduction to batteries*

A battery is a stack of electrochemical cells connected in series and/or parallel. A cell is an electrochemical device that transforms chemical energy into electrical energy and then electrical energy back into chemical energy [10]. Cells can be classified as either primary or secondary cells. Primary cells are non-rechargeable while secondary cells are rechargeable. This study was based on secondary cells as they are used in this design. The electrochemical cell as shown in Figure 2-4, consists of a negative electrode, a positive electrode, an electrolyte solution containing dissolved salts and a semi-permeable membrane.

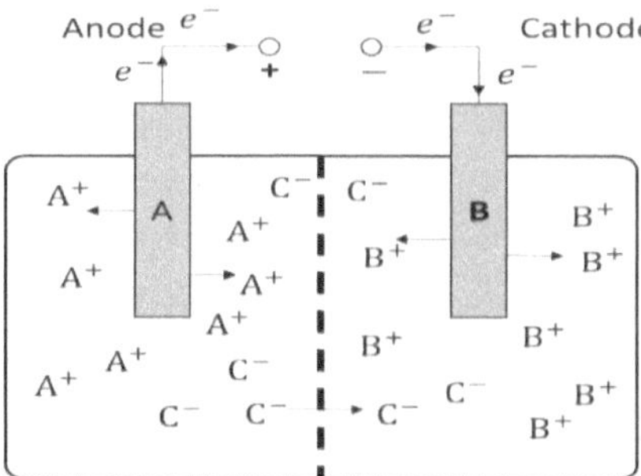


**Figure 2-4 Secondary cell in charging mode**

The negatively charged C ions are the salt solution and the bold dotted line is the separator. Only the salt solution ions can pass through the semi-permeable membrane. At the anode, Metal A gets oxidised and transfers its valence electrons to the other metal. Once it transfers its electrons, its ions get discharged into the solution. At the cathode the opposite happens, the metal B gets reduced and gains the electrons lost by A. This results in an electrical current flowing.

## 2.2.2    *Electrical introduction to batteries*

A battery provides DC electrical power to a circuit. It is only capably of direct current which is current in one direction. Figure 2-5 shows one of many electrical models of a battery.

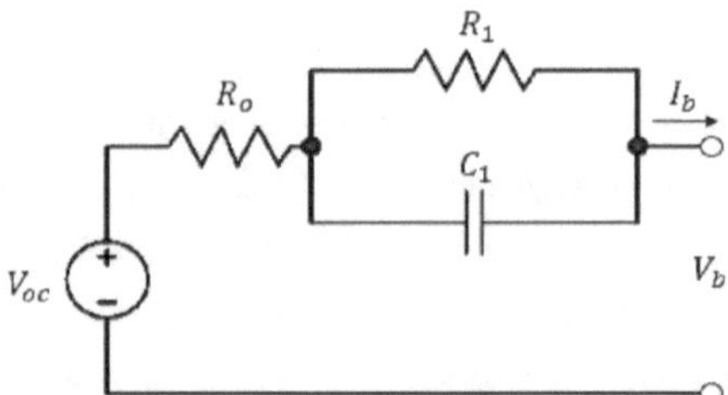

**Figure 2-5 First order battery electrical equivalent circuit**

$V_{OC}$ represents the open circuit voltage of a battery, $R_0$ is the ohmic resistance of the battery associated with the electrolyte resistance and battery connection resistance. $R_1$ represents the polarization resistance and $C_1$ represents the polarization capacitance [11]. $V_b$ in this case would be the terminal voltage of a battery. One ampere is exactly 1 coulomb of electrons per second [12]. All batteries are rated in amphours (Ah). A battery that is rated 1 amp-hour will, ideally be able to provide 1 ampere to a complete circuit for one hour before completely discharging but in real batteries the relationship between the amperes and discharge time is not linear. A battery C rating of amp-hour capacity is provided. A battery also has watt-hour ratings which is the total constant DC power provided by a battery before it discharges completely. A battery rated at 1 watt-hour can provide 1 watt of constant power for 1 hour before it discharges completely [12]. Some cells suffer from what is called the Memory effect. This is a phenomenon whereby if a cell isn't discharged fully between charge cycles, it starts adapting to the previous shortened cycles and thus reduces the capacity of the cell per charge cycle [13].

### 2.2.3    *Charging a battery*

Charging a battery is a significant part of any power supply design. The charging system depends on the battery being charged. There are two ways of charging a battery – slow and fast charge [14]. Slow charging a battery refers to applying a current to a battery for a long period of time without damaging the battery. This charging rate depends on the type of battery used. If a battery is charged continually even after it's fully charged, gas begins to form in the battery. For slow charging, the gas can recombine internally but if charging rate is above slow charging rate, the gas is unable to recombine, and pressure builds up inside and damages the battery in turn reducing the battery life cycle. The big advantage of slow charging is that it cannot damage the battery regardless of how long the battery is charged. The other advantage is that no end-of-charge detection circuitry is required meaning it's cheap and simple to build [14].

Fast charging refers to a charging time of between 1- and 2-hours max. Fast charging usually occurs when the cell's temperature is between 10 and 40 degrees Celsius. For temperatures below 20 degrees, the gas builds up raises the pressure more quickly and can damage the battery, it is therefore advisable to operate at 25 degrees Celsius. The typical charging curve of a battery is shown in Figure 2-6. This one refers to a typical lithium-ion battery charging curve. Several batteries have different charging curves due to their chemical composition and structure [14].

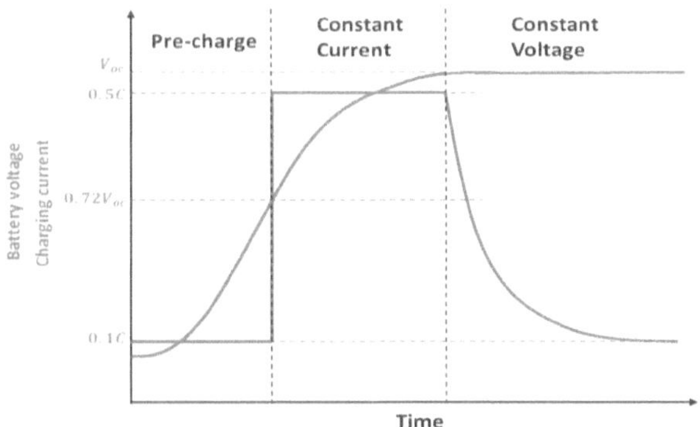


**Figure 2-6 Lithium-ion cell charging curve**

The charge curve in Figure 2-6 consists of 3 charging stages: pre-charge; constant current (CC) and constant voltage (CV) stages [15]. If the cell voltage is below 10% of the full open circuit voltage, the charge current used is termed the pre-charge current to prevent damage to the cell. This pre-charge is applied to the cell until the cell voltage reaches a threshold voltage of approximately 70% of the open circuit voltage. After that stage, constant current is applied. Cells are usually charged at 0.5C or less i.e. a charging current half of the cell's rated current. Once the voltage of the cell reaches the max voltage of 4.1V, the current reduces. There are several types of battery charging methods. A few were reviewed.

*i.*     ***Constant voltage***

A DC power supply with a step-down transformer and a rectifier to provide constant DC voltage. To protect against battery damage, regulatory circuitry is included in the charger [16].

*ii.*        **Constant Current**

The charger is based on an unregulated voltage source. As shown in Figure 2-6 the current decreases as the battery voltage builds up. Often, protection circuitry is included to prevent overcharging. This type of charge is not suitable for all batteries [16].

*iii.*        **Trickle charge**

Designed to account for the battery self-discharge. In this type of charging, the charge rate differs depending on the frequency of discharge. It is not suitable for batteries that are vulnerable to overcharging like Li-Ion Batteries [16].

*iv.*        **Float charge**

This type of charge involves the battery and load being connected in parallel across the charger. They are held at a voltage level below the battery voltage charging limit. This type of charging is used mainly in back-up power systems. Lead-Acid batteries use this type of charging [16].

### 2.2.4    *Discharging a battery*

The discharge voltage appearing at the terminals of the battery depends on the type of current that the load draws; the internal resistance of the battery which varies with temperature, state of charge of the battery and age of the cell [17]. Figure 2-7 shows typical discharge curves of several cells.

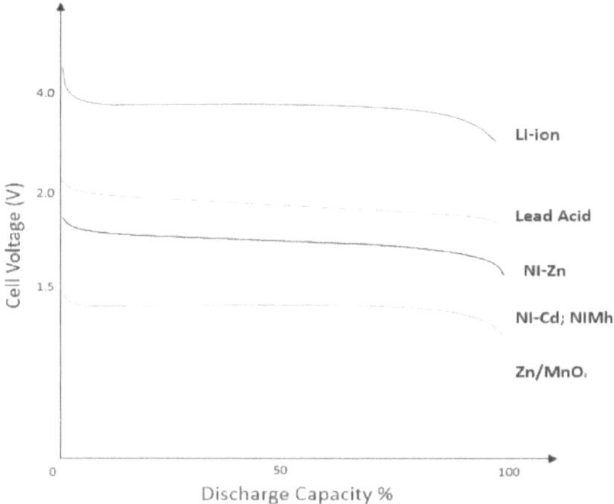



**Figure 2-7 battery discharge curves of two batteries at a rate of 0.2C**

Of all the types of batteries, lithium-ion battery has the highest open circuit voltage and its discharge curve is flat. The lead acid battery has the next highest open circuit voltage.

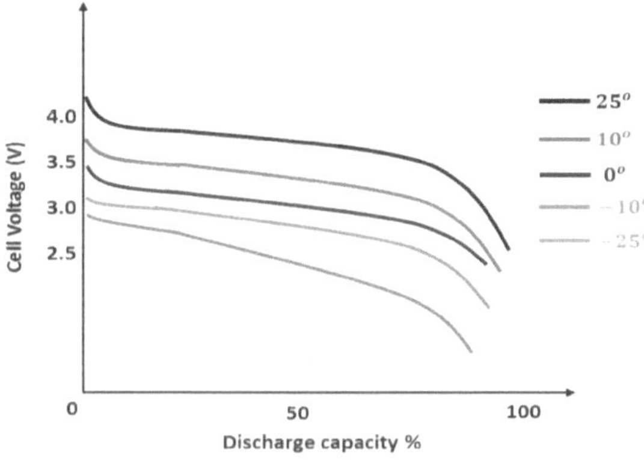



**Figure 2-8 Lithium ion battery discharge curves at varying temperature vs discharge time**

Figure 2-8 shows that the lithium-ion battery performs better at temperatures above 20 degrees. At low temperatures the electrolyte may freeze and suffer from lithium plating at the porous carbon electrode. At higher temperatures the battery may get damaged. Self-discharge is another factor of a battery discharge where it discharges on its own without being used. Lithium-ion batteries have lowest self-discharge rate of all batteries at 2% per month and nickel metal hydride batteries have the highest at 30%

per month. Lead acid batteries self-discharge at 4% per month. At temperatures below 20 degrees the self-discharge rate is at its lowest and gets higher as the temperature increases. Internal resistance is another factor to consider when discharging a battery. It decreases the terminal voltage of the battery and increases the voltage needed to charge the battery. Figure 2-9 shows the discharge curves of a lithium-ion batteries at different C rates [17].

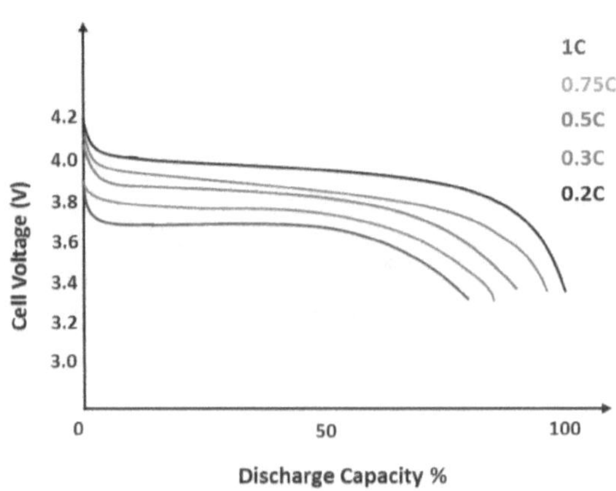


Figure 2-9 Lithium ion battery discharge curves at different C rates

The higher the discharge rate, the higher the voltage drop. A battery can be discharged at different C rates depending on the load. At 1C, the battery discharge curve is flat and has a high open circuit voltage. The higher the discharge rate, the lower the discharge voltage.

### 2.2.5 Types of secondary batteries

The various types of secondary cells are listed and discussed.

### i. Nickel-Cadmium

This type of battery uses nickel oxide hydroxide and metallic cadmium as electrodes. The cells use potassium hydroxide as an electrolyte [18]. They have low internal resistance meaning they can supply high current without overheating. To deliver the high current, the anode and cathode are rolled into a spiral. The nominal voltage of the cell is 1.2V. the energy density of the cell is 50-150 W-h/L. It can be recharged 2000 times in its life-span. Its specific power is approximately 150W/kg at a discharge efficiency of 70-

90%. The advantages of the battery are that they are cheaper than Li-ion batteries, they have a fast charge cycle and low temperatures do not affect the battery. The disadvantages of the battery are that they have a low energy density and they also are not environmentally friendly, since they contain toxic metals [19].

## ii.      *Nickel-Metal hydride*

This type of battery is like the previously mentioned nickel-cadmium battery because they both use nickel oxide hydroxide as the cathode, but the nickel-metal hydride battery uses a hydrogen-absorbing alloy for the anode instead of cadmium [20]. The separator in then made up of fibres, an alkaline electrolyte, a metal case and a sealing plate. They can also be arranged into a spiral like the Ni-Cd batteries. It also has a nominal voltage of 1.2V with an energy density of 140-300 W-h/L. It also can be recharged about 2000 times with a charge-discharge efficiency of 66-92%. The advantages of the battery are that is environmentally friendly, and it makes recycling profitable. They do not require any circuitry to regulate the power like the lithium-ion batteries. The disadvantages are that they require complex charging circuitry, they have a high self-discharge and they generate a lot of heat [16].

## iii.      *Lead acid*

At the anode of a lead acid battery is an electrode made up of pasted lead arranged into a grid. At the cathode is an electrode made of lead dioxide. The two electrodes are submerged in a sulphuric acid electrolyte. The lead ions in the electrolyte combine with the sulphate radicals to produce lead-sulphate plus an electron.t this causes an electrical current to flow [21]. The lead acid cell produces a nominal voltage of 2.2V during discharge. This battery has the lowest energy to weight ratio. The most widely used types of lead acid batteries are sealed and flooded. Because of the acid in the cells, these cells are not environmentally friendly.

## iv.      *Lithium-ion*

Lithium is the lightest of all metals with the greatest electrochemical potential and provides the largest energy per kg [21]. The battery uses metal oxide as the positive electrode and porous carbon as the negative electrode. During charging, electrons flow from the cathode to the anode. The battery has an energy density of 250-693 Watt-h/L and an open circuit voltage of 4.2 V and a terminal voltage of 3.7V. Its charge efficiency is 80 to 90%. One of the main advantages of the lithium-ion battery is its high energy density compared to other batteries. There is no need to prime this type of battery. It is limited, however because it requires circuitry to keep the voltage within its safe limits. They also suffer from ageing and they have certain travel restrictions such as not being able to carry more than needed on an aircraft. They also cost significantly higher than the other batteries [21].

## 2.3    Low voltage grid-tied bidirectional power conditioning system (PCS) architectures

Lithium-ion batteries are best solution for storage because of the advantages already reviewed. As such, they were chosen as the primary power source for the PCS to be designed. Residential AC loads use single phase line voltage – as such, only single-phase PCS's were reviewed. PCS's can be divided into two categories – voltage source inverters (VSI's) and current source inverters (CSI's) [22]. CSI's use a DC current source as the input and are typically used in powerful AC motor drives. VSI's are the most commonly used type of inverter. They can either be transformer-based or transformer-less inverters. These inverters were the main type of PCS's reviewed.

### 2.3.1    *Transformer-based bidirectional inverters*

Some inverters with a front-end DC-DC converter, use transformers for electrical isolation but topologies without a DC-DC converter use the transformer for both step-up or step-down purposes and electrical isolation. Transformer based inverters use either a high-frequency of low-frequency transformer. These were reviewed.

#### *i.    High-frequency transformer bidirectional inverters*

Figure 2-10 shows a block diagram for a two-stage high frequency (HF) transformer-based inverter using batteries as an input DC voltage source.

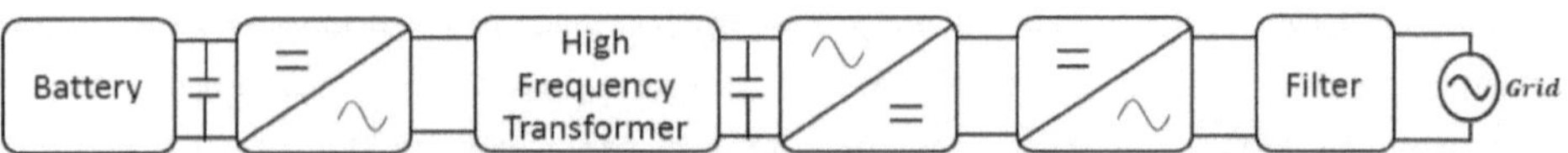


**Figure 2-10 HF transformer inverter block diagram**

The high frequency transformer is connected between the power stages of the system. A high frequency transformer requires these two-stage shown above, as such the efficiency of the HF transformer is compromised [23]. Some topologies based on the block diagram were reviewed.

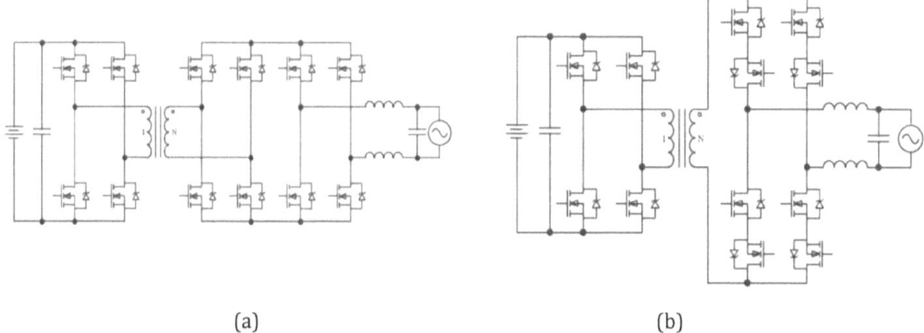

(a)      (b)

**Figure 2-11 HF link inverter topologies (a)  Bidirectional multiple-stage boost inverter (b)  Bidirectional dc–ac–ac converter**

The topologies shown in Figure 2-11 all have bidirectional capabilities. They all have at least two power stages, this is because the HF transformer requires AC voltage on either side. The HF transformer provides electrical isolation for safety and protection. Figure 2-11 (a) shows a DC-AC-DC-AC topology with DC-DC pseudo-DC-link stages. This stage features DC pulse trains with a period that is half that of the fundamental AC output. To ensure a low THD at the output, a low pass filter is added. It can buck during DC-AC power flow and boost during AC-DC power flow. The last stage is a voltage source. Figure 2-11 (b) shows a DC-AC-AC topology. The last stage is an AC-AC stage with a DC-link to eliminate the bulky DC-link capacitor from multiple-stage topologies [24].

*ii.*  ***Low-Frequency transformer bidirectional inverters***

Low frequency transformers are connected on the grid-side of the inverter unlike the HF transformers discussed above. The block diagram for these topologies is shown in Figure 2-12.

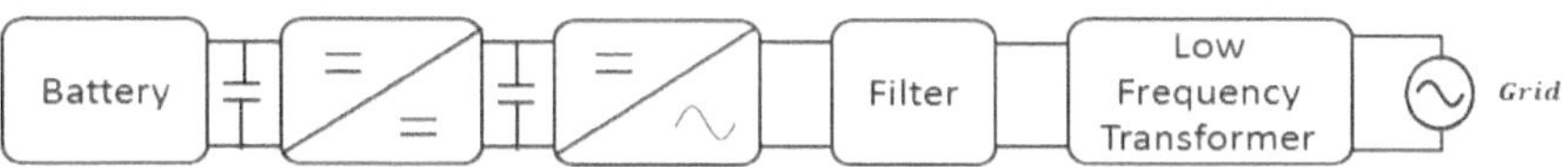


**Figure 2-12 LF transformer inverters block diagram**

These inverters do not require as many power stages as shown in the block diagram. But they are larger, heavier and have a larger cost than HF transformers [23]. These inverters also have a low efficiency associated with them.

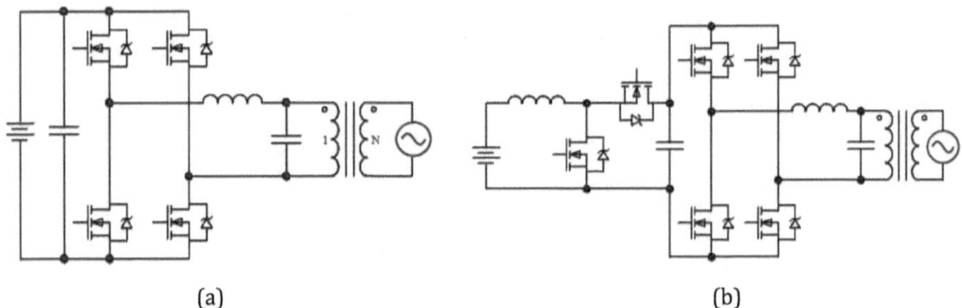

(a)                                          (b)

**Figure 2-13 Low frequency inverter topologies (a) Bidirectional isolated buck inverter (b) Bidirectional two-stage isolated buck boost inverter for PV applications**

Figure 2-13 shows that these inverters do not require as many power stages as the HF based inverters. Figure 2-13 (a) shows a traditional bidirectional buck inverter that uses an Unfolding bridge and bidirectional switches. It steps down voltage both from DC-AC and AC-DC. A low pass filter is then used to filter out the high frequency ripple. The transformer then provides the isolation between the grid and circuit. Figure 2-13 (b) shows a bidirectional inverter that boosts during AC-DC power flow and bucks during AC-DC power flow. It is suitable for low voltage applications. But the issue with low frequency transformers is that they are large and because of this tend to have a lower efficiency that the HF transformers [23].

### iii.     *Transformer-based bidirectional inverters comparison*

The inverter topologies discussed in the previous section were reviewed in terms of their efficiency, weight and volume.

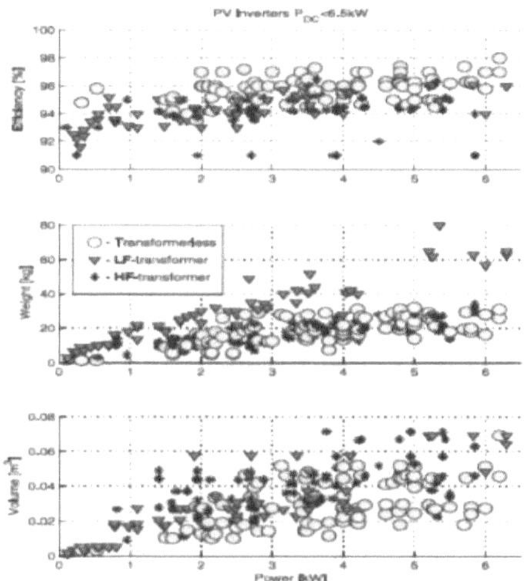


**Figure 2-14 Transformer inverter topologies comparison [25]**

Figure 2-14 shows that the efficiency in transformer-less inverters is highest overall as reviewed in the previous section. The low frequency transformer has the next highest efficiency and the HF transformer has the lowest. The low frequency transformer is shown to have the heaviest topologies and transformer-less the lightest. Transformer-less inverters clearly have an advantage over the other inverters and are generally preferred [25].

## 2.3.2 *Transformer-less bidirectional inverters*

From the in the previous subsection, it was clear that transformer less inverters have an advantage of the transformer-based inverters. As such, the design undertaken was based on transformer-less inverters. The several architectures associated with transformer-less inverters were reviewed.

### i. *Two-stage topologies*

Inverters that are based on two power conversion stages were reviewed and the block diagram for typical two-stage topologies is shown in Figure 2-15.

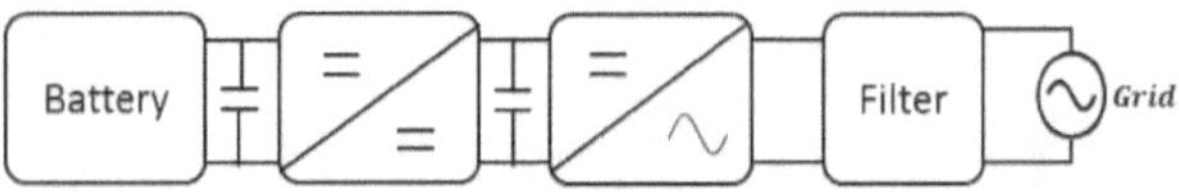

Figure 2-15 Two-stage transformer-less Inverter block-diagram [12]

The first stage is a step-up stage where the DC input voltage is stepped up, the next stage is then when the DC input is inverted into AC using an Unfolding bridge and then connected to the Grid. During the DC-DC stage, the inverter can be designed with any DC-DC converter topology. Figure 2-16 shows a bidirectional buck-boost inverter.

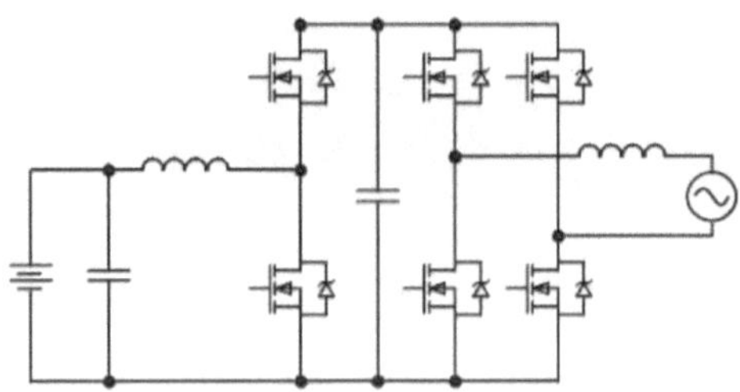

Figure 2-16 Two stage buck-boost bidirectional inverter topologies

During AC-DC power flow, the inverter acts as a boost inverter. Therefore, the AC voltage is always lower than the DC voltage of the battery. This type of bidirectional inverter is mostly used as an interface between the grid and electrical vehicles [26]. Figure 2-17 shows an integrated bidirectional inverter.

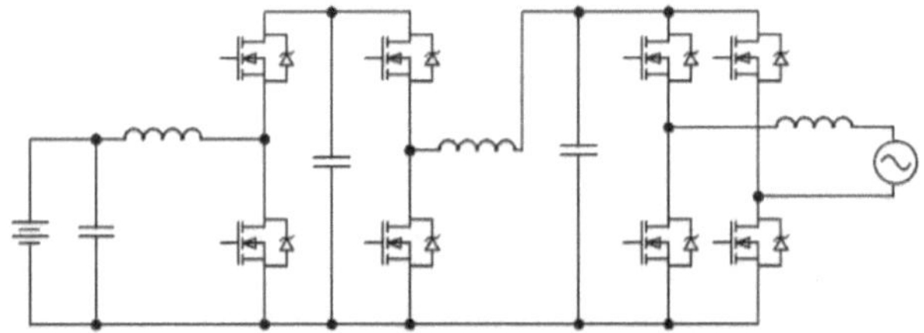

Figure 2-17 Two stage integrated bidirectional buck boost inverter

Unlike the topology shown in Figure 2-16, the integrated buck boost inverter can either buck or boost voltage during DC-AC and during AC-DC power flows. This allows better for management of the battery power. But because it has more components than the topology in Figure 2-16, it has a lower efficiency [26].

The advantages of this type of architecture are that it has a wide input voltage range and the dynamic analysis is easier. The drawback of this topology is that it requires more components since it has two power stages. It also requires a DC-link capacitor to stabilise the DC input voltage of the inverter side which draws a large current which means an expensive filter is required at the output.

*ii.* **Single-stage topologies**

To overcome the drawbacks of the two-stage topologies, single stage inverters with a block diagram as shown in Figure 2-18. The DC input power is converted into AC power during one power conversion stage.

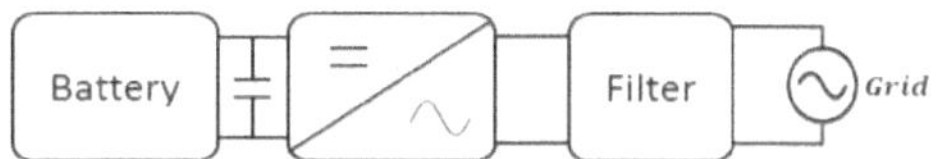


**Figure 2-18 Single stage inverter block diagram [12]**

The base DC-DC converters that have been used with this configuration are any of the common topologies; this is because of their simple structure and low component count.
The reduction in power stage means that the efficiency is higher; the cost of the inverter is reduced; the inverter is compact, and the output filter design is cheaper since there is no DC-link capacitor. The disadvantages of single-stage topologies are that, they have a narrow input voltage range; they require complex control design; and because of their compact structure, the existing step up bidirectional inverters use a HF-link.

*iii.* **Pseudo-dc-link topologies**

The frond end DC-DC converter produces a modulus sinusoid during the first stage, the second stage is an unfolding bridge which is used to unravel the modulus sinusoid and produce the required bidirectional sinusoidal voltage. These types of inverters are between the two-stage and the single-stage in terms of efficiency, weight and volume, assuming they are designed without a DC-link capacitor. They can have the efficiency profile similar to single stage inverters. The cost is also lower than that of the two-stage topology. The block diagram for these types of topologies is shown in Figure 2-19.

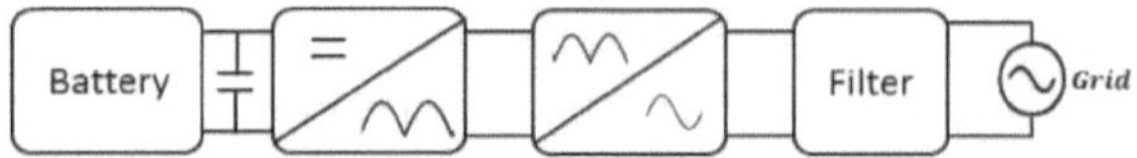

Figure 2-19 Pseudo-dc-link inverter block diagram [12]

Because the rectified sinusoid is produced during the first stage of power conversion, only buck-boost, Ćuk, Sepic and zeta DC-DC converters can be used for pseudo-dc-link inverters. Bidirectional pseudo-dc-link inverter topologies also exist – but only the Ćuk and buck-boost topologies are capable of bidirectional operation because of the symmetry of the converters. The most popular pseudo-dc-link topologies are buck-boost because of the low component count and ease of design. Figure 2-20 shows a buck-boost bidirectional pseudo-dc-link topology.

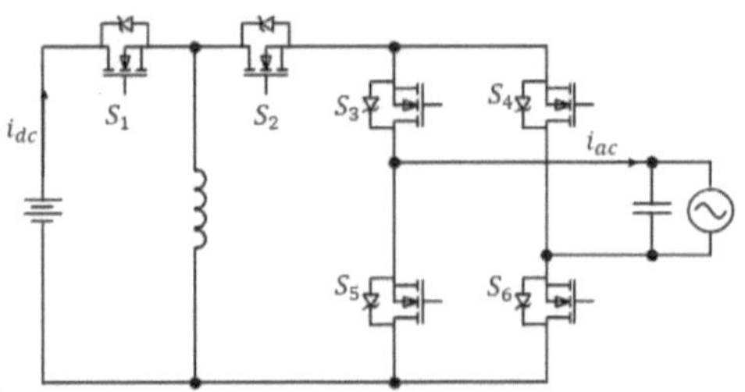

Figure 2-20 Bidirectional pseudo-dc-link buck-boost inverter

The inverter uses a front-end buck-boost converter and an unfolding bridge to create the line frequency AC output voltage. During DC-AC power flow, it can either buck or boost depending on the application and the same applies during AC-DC power flow. The buck boost switches $S_1$ and $S_2$ switch at 20 $kHz$ while the unfolding bridge switches are switched slower at the fundamental grid frequency to create the AC line voltage [27]. This inverter is compact and low cost since it was designed without a DC-link capacitor. Figure 2-21 shows another bidirectional pseudo-dc-link inverter.

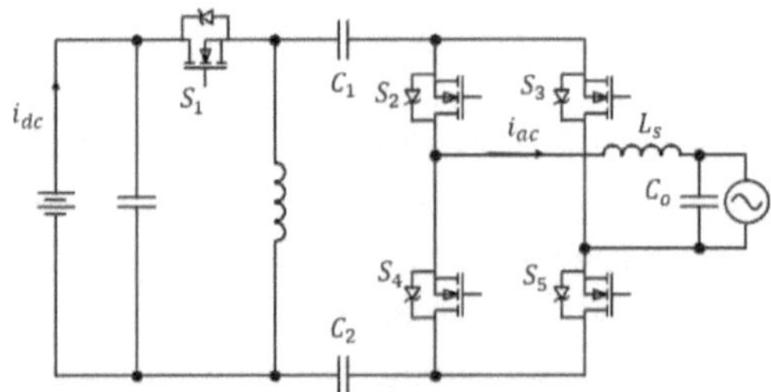

Figure 2-21 Bidirectional pseudo-dc-link Sepic-Zeta inverter [28]

During AC-DC power flow, the inverter operates as a Sepic AC-DC converter [28]; if operated with both inductors in discontinuous inductor current mode, it can achieve self-power correction. During DC-AC power flow, it operates as a Zeta inverter. It can buck or boost during either power flow, depending on the application. Its key features are low cost, high efficiency and provides the best of two converters. The drawback to this topology is that, two converter topologies have to be studied and analysed for the inverter to be designed adequately because neither Sepic nor Zeta have the symmetry for bidirectional capability.

Although the buck-boost based topology is the most popular over Sepic, Ćuk and Zeta, the buck-boost has both input and output currents that are discontinuous, as such for practical implementations, filters at both input and output are required to reduce the line current ripple and EMI. The Ćuk topology is then best for the practical applications because it has both continuous input and output currents; providing galvanic isolation is easier and during reverse power flow rectification, it has a better input line current than the buck-boost.


## 2.4    Control of Ćuk topologies

There are several methods of control that have been implemented to ensure robustness in Ćuk topologies. Current and sliding mode control are the most widely used and were reviewed for possible implementation.


### 2.4.1    *Sliding mode control (SMC)*

Sliding mode is a variable control system that can switch between continuous structures based on the system's position in a state. It is used to control non-linear systems; multiple input and multiple output (MIMO) systems; and dynamically uncertain systems [29]. Figure 2-22 shows a simplified block diagram of SMC. It has a set valued, discontinuous control signal that can switch from one continuous structure to the next by forcing the system states to reach and remain within what is called a siding surface.

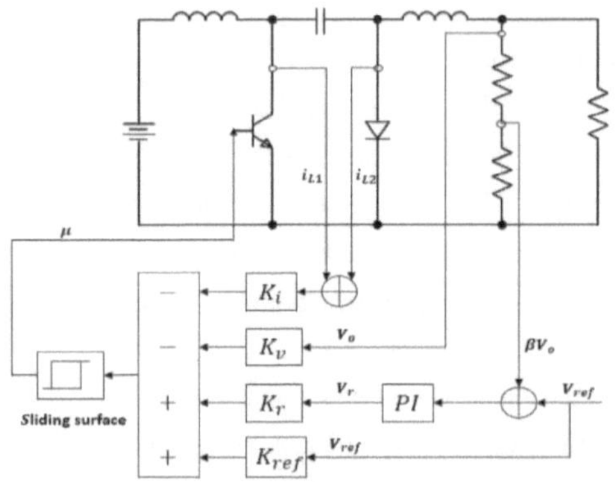



**Figure 2-22 Sliding mode control block diagram for a Ćuk converter**

A control law is designed to ensure the sliding surface attracts system states. When the system dynamics are confined to the sliding surface, the order of the original plant, as seen by the system, is reduced. When the order of the plant is reduced, the closed loop system becomes insensitive to plant uncertainties and disturbances. Another attractive feature of the SMC is that the dynamic behaviour can be tailored to a sliding function. In [30] a sliding mode controller was designed for a 3rd order Ćuk converter. This SMC used 3 sliding surfaces to control the output of the Ćuk converter. The results revealed that the higher the number of sliding surfaces, the better the performance of the output voltage. The results showed that although, the performance increased with sliding surfaces, there was still a significant overshoot during the load disturbance tests – the voltage had an overshoot of more than 30% which is not ideal for long term use of batteries which can only handle high voltages for a short time. In [31] a sliding mode controller was designed for Ćuk based solar inverter. Only a single reference state was used. The advantage of this control is that, no high pass filters were needed to derive other state reference signals. The results also showed larger overshoots during load disturbance tests which would also not be ideal for long term use of batteries.

However, to force the system into states that will make it slide along the sliding surface, infinitely high frequency is required, as reviewed. But it is not possible to switch the control system infinitely fast between continuous structures because there will be time delays between components that realise the control system such as the active power switches, transducers, DSP units and software implementations. This limitation leads to high frequency oscillations, high power losses and could cause the system to move into an unstable state. Some SMC can also fail to reject un-modelled uncertainties which compromises the performance of the PCS.

## 2.4.2    *Current control*

In PCS's, voltage control uses the difference between the reference signal and the sensed output voltage to produce an error signal which compensated for by a controller and the output is the duty cycle control input signal which is compared with a high frequency carrier signal to produce the necessary switching signals. Current control however, is type of control scheme that uses two loops to control the output voltage. The error signal from the difference between the reference and sensed output voltage, goes to a controller after which, the control signal is compared to the inductor current. This current comprises of a high frequency AC ripple component and an average current component, instead of the high frequency carrier signal. The output then produces the necessary switching signals [32].

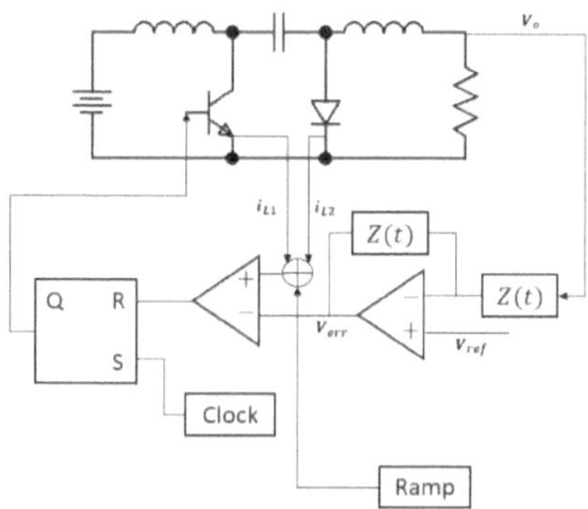


**Figure 2-23 Peak current mode control block diagram for a Ćuk converter**

There are 3 types of current control methods: peak, average and valley. Peak mode control is the most widely used because of the easier implementation. But since the peak current is being controlled, at zero crossings, the slope $\frac{di}{dt}$ is slow due the large inductance used for continuous inductor current modes. This leads to more crossover distortions, since the error voltage is small and there is a mismatch between the peak current and the error signal. To solve this issue, a ramp compensator is designed. This ramp is added to the peak current of the inductor to decrease the mismatch and in so doing, stabilises the closed loop. But the ramp limits the duty cycle at zero crossings. This leads to the peak mode having low gain in the current loop and high bandwidth. As such, for AC-DC power flow, this type would not be suitable. Average mode current control used to solve the drawbacks of peak mode. The average inductor current is controlled which eliminates the need for ramp compensation since there is no mismatch between the error and ripple current. To solve the low gain issues of peak current mode, a high gain integrating factor is introduced into the current loop. This results in low noise and low harmonic distortion at the output. Although slope compensation is not required, the current loop gain is limited [33]. Valley current mode is rarely used because of the success of the average mode control. In [34] an indirect current control was implanted for a Ćuk converter with a NI-Mh battery input for energy harvesting. Instead of a current sensor, the output voltage of thermo-electrical generators was to indirectly measure the current. The results showed adequate output voltages with significant overshoots and some output voltage variances during load regulation. In [35] peak current mode was employed with a ramp compensator to improve the stability of the converter. The results showed a closed loop with a narrow stability margin as the duty cycle is above 0.5. This is not ideal for low voltage power applications since high duty cycle values are used.

## 2.5    Summary of reviewed literature

Based on the literature done, the best set up for grid-tied residential low voltage is one with a storage solution over a set up with no storage solution due to advantages reviewed in 2.1.2. The best storage device is a lithium ion battery bank over other reviewed storage devices and other batteries. Transformer-less inverters are preferred over transformer-based inverters due to advantages discussed in 2.3.1iii. From transformer-less inverters, even though two-stage topologies have the widest input voltage range and easier dynamic analysis, they are less efficient and have a higher cost than single-stage topologies. Although single-stage topologies are better than two-stage, their compact structure make it difficult to configure them for step-up bidirectional operation – as required for low voltage battery applications. Pseudo-dc-link topologies are less efficient than single-stage topologies but are more efficient than two-stage due to the elimination of the DC-link capacitor – also making them less costly. Because of their structure they can have an efficiency close to single-stage topologies but – unlike single-stage topologies – have bidirectional capability. This made them the best choice for a bidirectional inverter design that is also efficient and has a minimal component count. But for pseudo-dc-link topologies, only Ćuk, Sepic, buck-boost and zeta converters can be used. Of those four, only the buck-boost and the Ćuk have the symmetry for bidirectional capability. But because of the buck-boost converter's discontinuous input and output currents and difficulty in isolating the converter, the Ćuk was chosen because it doesn't have the drawbacks of the buck-boost. Because of the drawbacks of the two control methods reviewed, a new control strategy was presented in this book.

# 3.  Bidirectional Ćuk inverter analysis

The theory development, operating principle and analysis of the proposed bidirectional inverter was presented in this section. The Ćuk converter – as the base DC-DC converter – had its power transfer analysed. The topology's operating principle was then analysed in both AC-DC and DC-AC power flows. In each power direction of power flow, the topology was analysed in each switching state. Then the modes of operation were analysed starting with CICM and DCVM. Dead time effects and parasitic effects were analysed after which the expected efficiency was analysed. Lastly, the dynamics of the topology was analysed.

## 3.1    Proposed topology overview

The bidirectional Inverter presented in Figure 3-1 is a pseudo-dc-link, Ćuk front-end single-phase inverter. It has 2 inductors, an input inductor $L_1$ and output inductor $L_2$; a coupling capacitor $C_1$ and filter a capacitor $C_2$; six switches, 2 during the converter stage and 4 in the unfolding stage. The inductor $L_s$ is the grid source inductance which was chosen to be 5% of the rated AC voltage.

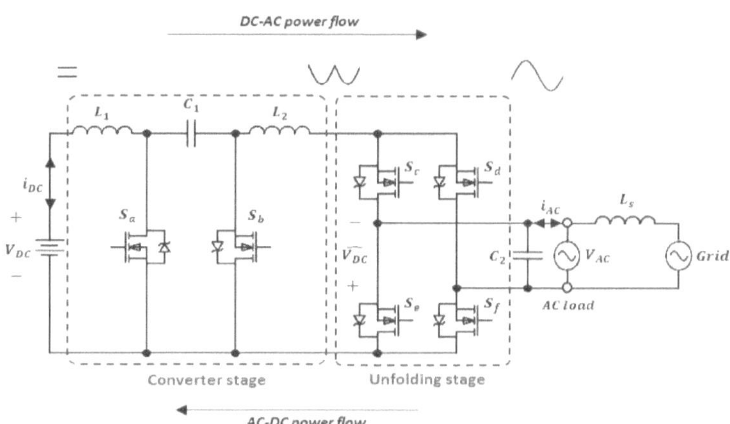


**Figure 3-1 Single-phase bi-directional Ćuk inverter**

Since a Ćuk converter is used, it can buck and boost during either DC-AC power flow or during AC-DC power flow. During DC-AC power flow, it is supplied from a DC input voltage, such as a battery. The coupling capacitor acts as the energy storage for the power stage. The duty cycle varies through switches, $S_a$

and $S_b$ – which are switched at $20kHz$ carrier frequency, $f_s$ – to produce a pseudo DC voltage, $\widehat{V_{DC}}$ at the output of the converter stage. This pseudo DC voltage is then cascaded with an unfolding bridge – which is slow switched at $50Hz$ fundamental grid frequency, $f_h$ – to produce the required AC load sinusoidal voltage. During reverse power flow AC-DC, the body diodes of the unfolding bridge switches, $S_c$ $S_d$ $S_e$ $S_f$, provide natural regenerative capabilities. Under ideal conditions, the power factor from the AC side is $(-1)$; this enables the power to flow from the grid back to recharge the battery – this will be shown in more detail in later sections. The AC grid voltage is given by (3.1) as:

$$V_{AC} = \left(\frac{d_{max}}{1 - d_{max}}\right) V_{DC} \sin(2\pi f_h t) \tag{3.1}$$

Where $d_{max}$ is the duty cycle required to reach the grid voltage peak from the battery terminal voltage; $f_h$ the fundamental grid frequency and:

$$\left(\frac{d_{max}}{1 - d_{max}}\right) V_{DC} = V_m \tag{3.2}$$

The voltage $V_m$ is the peak grid voltage. The inverter offers advantages of being compact and can step up low voltage and step down the grid voltage without sacrificing the size of the system. It is low cost and highly efficient, as will be analysed in later sections. It also has a high frequency link option, which is not an option with other front-end DC-DC converters.


## 3.2    Bidirectional inverter operational analysis

The bidirectional inverter operation during DC-AC and AC-DC power flows was analysed. The equivalent circuits were derived for each power flow and the conducting patterns analysed. After which, the mathematical analysis of the AC and DC output voltage productions was reviewed.


### 3.2.1    *Inverter modes of operation*

During DC-AC power flow, the power factor at AC side is one – under ideal conditions. This enables the power to flow from the battery to the AC load. During AC-DC power flow the power factor is negative one – this enables the power to flow from the AC side back to the battery. Figure 3-2 shows the modes of operation for the inverter.

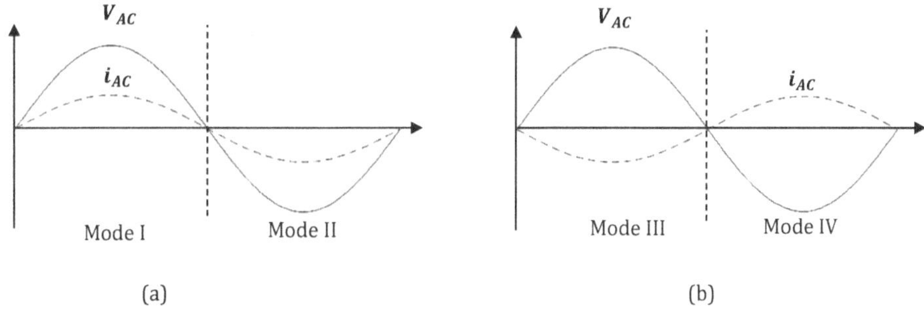



**Figure 3-2 Bidirectional inverter modes of operation at an ideal power factor (a) DC-AC power flow (b) AC-DC power flow**

Within a critical loads panel in a residence are linear resistive loads like incandescent light bulbs and non-linear resistive loads like LED or plasma TV's. If the inverter is assumed to have a purely resistive AC load, then Figure 3-2 (a) shows that the AC voltage and current are in phase; this gives positive power which flows from the battery to the AC side. Mode I occurs when both AC current and voltage are positive. Mode II occurs when both AC current and voltage are negative. Figure 3-2 (b) shows the voltage and current are $180^{o}$ output of phase; this gives negative power which flows from the AC side back to the battery. With this load, the inverter also only has two modes. Mode III occurs when the AC voltage is positive and AC current is negative. Mode IV occurs when the AC voltage is negative, and the AC current is positive. Figure 3-3 shows cases when the load is either capacitive or inductive without reactive power compensation.

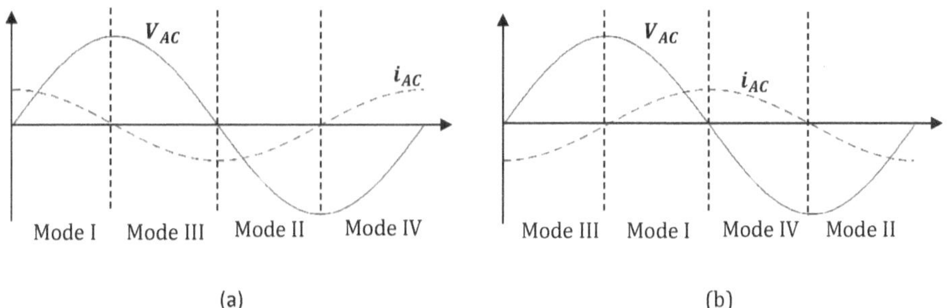



**Figure 3-3 Bidirectional inverter with non-ideal power factor (a) Leading power factor (b) Lagging power factor**

The critical loads panel might also have capacitive loads like old CTR TV's. Inductive loads like ceiling fans and any other household appliances using motors or transformers in them. When a capacitive load is used, the power factor is leading; the AC current is shown to lead the AC voltage; the inverter then has 4 modes of operation. With an inductive load, the power factor is lagging; the AC current lags the voltage, and the inverter also operates in 4 modes.

### 3.2.2 DC-AC power flow inverter equivalent circuit and operation

This section details a thorough analysis into how the inverter, with a resistive load, converts DC into AC power through Mode I and II. During DC-AC power flow, not all the components in the topology are used. The equivalent circuit for this is shown in Figure 3-4.

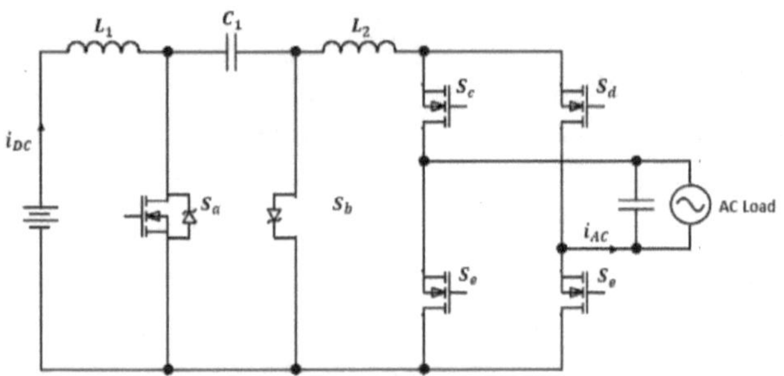

Figure 3-4 Ćuk inverter DC-AC equivalent circuit

The circuit only operates five of the six switches. Only the diode of switch $S_b$ is operational.

### i. Current analysis

Each half cycle creation was analysed individually, with each state and current direction shown.

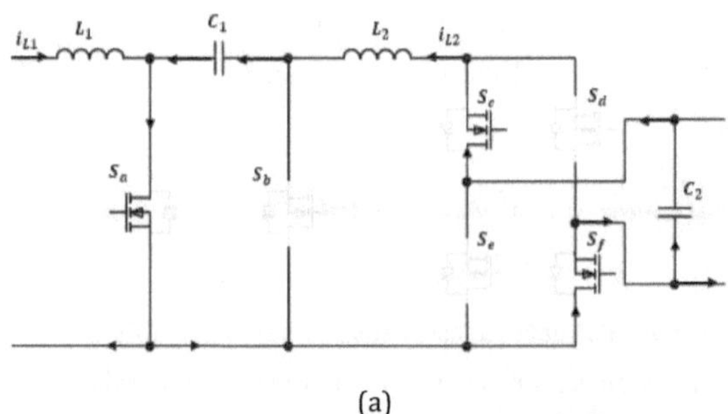

(a)

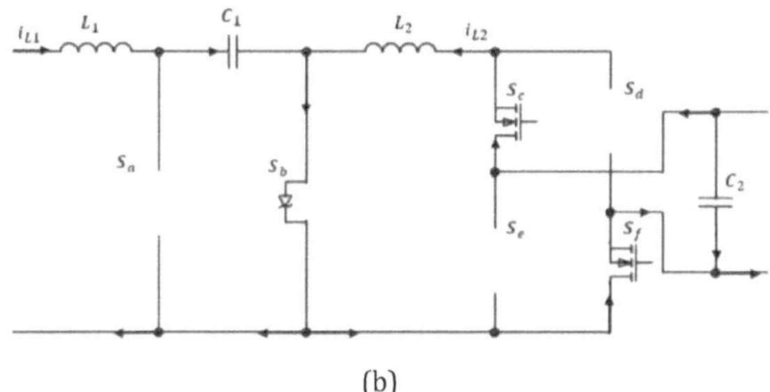

(b)

**Figure 3-5 Ćuk inverter DC-AC negative half cycle high frequency ripple (a) ON-state (b) OFF-state**

During the ON-state shown in Figure 3-5 (a), the input capacitor $C_1$ is discharging through the input inductor $L_1$ while the inductor charges through the input switch $S_a$. The diode of switch $S_b$ is reverse biased so current doesn't flow through it. In this state the inductor voltage is equal to the input voltage and the current is given by:

$$\frac{di_{L1}}{dt} = \frac{V_{DC}}{L_1} \tag{3.3}$$

In the unfolding stage, switches $S_c$ and $S_f$ are conducting and the output capacitor then charges up. The input capacitor $C_1$ transfers energy to the output capacitor $C_2$ for storage. The inductor $L_2$ charges from the output capacitor as it discharges, and its current is given by the equation (3.3) as:

$$\frac{di_{L2}}{dt} = -\frac{V_{c1} + \widehat{V_{DC}}}{L_2} \tag{3.4}$$

Equation (3.4) shows that the output inductor current is in the opposite direction to the input inductor current and charges with a larger voltage than the input inductor current. The OFF-state is also shown in Figure 3-5 (b) - During this cycle, the input inductor $L_1$ is discharging through the input capacitor $C_1$, which charges the input capacitor. The diode of $S_b$ is now forward biased. The inductor discharge current equation is shown in (3.5).

$$\frac{di_{L1}}{dt} = -\frac{V_{c1} - V_{DC}}{L_1} \tag{3.5}$$

The stage of the unfolding stage switches still haven't changed because the period of the fast switches is higher than those in the unfolding stage. Currently, the output capacitor $C_2$ is discharging through load resistor while the output inductor $L_2$ discharges through the diode of $S_b$. The equation through which the output inductor discharges is shown in (3.6):

$$\frac{di_{L2}}{dt} = \frac{\widehat{V_{DC}}}{L_2} \tag{3.6}$$

It is worth noting that, because the converter is symmetrical, power can be transferred from the right-hand side to the left-hand side and switch $S_b$ used as the primary switch. The same equations would apply as analysed above would apply – accept, the input would be on the right-hand and the output on the left. This alternation occurs for half the period of the bridge switches.

The positive half cycle shown in Figure 3-6 follows the same suit as the negative half cycle, having both ON and OFF-states for half the period of the bridge switches. The current analysis is shown in Figure 3-6 (a) and (b) for the charging and discharging states.

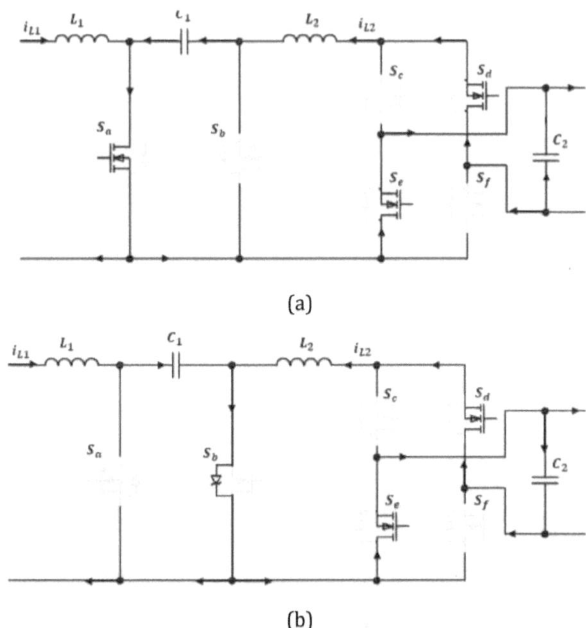



**Figure 3-6 Ćuk inverter DC-AC positive half cycle high frequency ripple (a) ON-state (b) OFF-state**

The ON-state operates similarly to the negative half cycle. The difference is in the bridge switch because only switches $S_d$ and $S_e$ are operational in this cycle. The OFF-state is also similar to that of the negative half cycle. The coupling capacitor waveforms are shown in Figure 3-7.

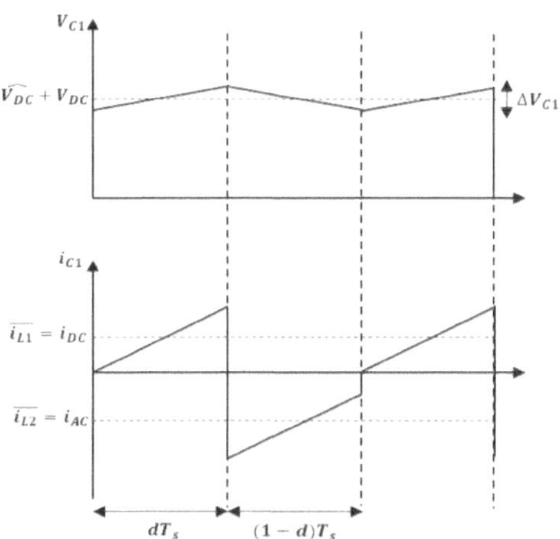

Figure 3-7 CICM coupling capacitor voltage and current

The coupling capacitor is assumed to be large enough such that, it has an average voltage that is the sum of the input voltage and the output voltage. The average current charging the capacitor is equal to the input current and the average value of the discharge current is equal to the output current. Figure 3-8 shows the waveforms of the inductors.

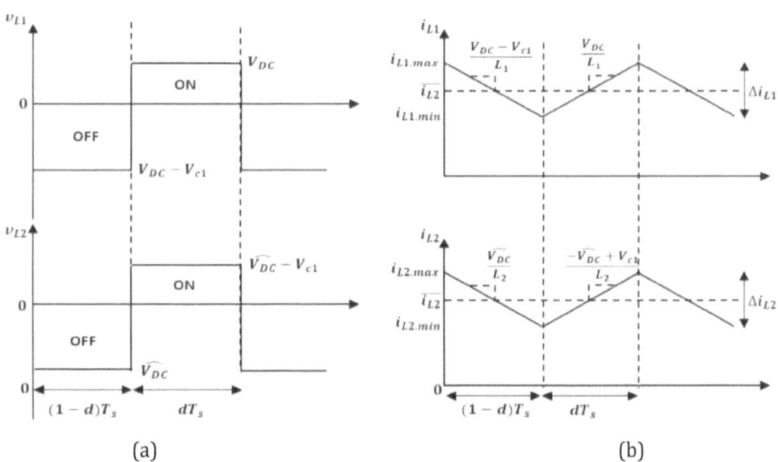

(a)          (b)

Figure 3-8 CICM input and output inductor waveforms (a) voltages (b) currents

The waveforms of the input and output inductor waveforms; both voltage and currents, are the same. The voltages do not differ in waveform or average value. In the OFF-state both inductors have a voltage equal to the output voltage and in the ON-state they are equal to the input voltage. The currents have the same waveform, but they differ in average values. The average value of the input inductor is the input current and average value of the output inductor is the output current.

*ii.*      **Continuous inductor current (CICM) ideal power transfer**

The inverter was designed such that during DC-AC power flow the inductors operate in continuous inductor current mode (CICM). During CICM, both inductor currents are continuous, and the input capacitor voltage is also continuous. If the voltage across the input capacitor is assumed to be constant, then the average value of the input and output inductor voltages at steady state is zero. Thus, for the input inductor:

$$\int_0^{T_s} V_{L1}\, dt = 0 \tag{3.7}$$

$$\int_0^{dT_s} V_{DC}\, dt + \int_{dT_s}^{T_s} (V_{DC} - V_{c1})\, dt = 0 \tag{3.8}$$

$$dV_{DC}T_s + (V_{DC} - V_{c1})(1 - d)T_s = 0 \tag{3.9}$$

where $d$ is the continuous-time averaged duty cycle, $T_s$ is the carrier period. Solving (3.9) for the capacitor voltage.

$$V_{c1} = \frac{V_{DC}}{1 - d} \tag{3.10}$$

For the output inductor:

$$\int_0^{T_s} V_{L2}\, dt = 0 \tag{3.11}$$

$$\int_0^{dT_s} (V_{c1} - \widehat{V_{DC}})\, dt - \int_{dT_s}^{T_s} \widehat{V_{DC}}\, dt = 0 \tag{3.12}$$

$$dT_s(V_{c1} - \widehat{V_{DC}}) - \widehat{V_{DC}}(1 - d)T_s = 0 \tag{3.13}$$

$$V_{c1} = \frac{\widehat{V_{DC}}}{d} \tag{3.14}$$

From equation (3.10) and (3.14):

$$\frac{\widehat{V_{DC}}}{V_{DC}} = \frac{d}{1 - d} \tag{3.15}$$

Equation (3.15) is the voltage transfer ratio during the converter stage of the inverter. Considering that the output voltage is an alternating sinusoidal voltage, the unfolding bridge can be described as the transfer function from the pseudo DC voltage to the AC output voltage:

$$U_b = \frac{|V_m \sin(2\pi f_h t)|}{V_m \sin(2\pi f_h t)} \tag{3.16}$$

This is because the Ćuk converter and the unfolding bridge are cascaded and are multiplied to obtain the output voltage. When this bridge function was plotted the following function was obtained. The unfolding bridge function can be described by (3.17) as:

$$U_b = \begin{cases} 1, & (n-1)T_h \leq t \leq \dfrac{(2n-1)T_h}{2} \\[4mm] -1, & \dfrac{(2n-1)T_h}{2} \leq t \leq n(T_h) \end{cases} \tag{3.17}$$

Where $n = 1,2,3 \ldots$ is a natural number; $T_h$ fundamental grid period. Therefore, the voltage transfer ratio from the input DC voltage to the output AC voltage after the unfolding stage – as a function of the duty cycle – is given by:

$$\frac{V_{AC}}{V_{DC}}(d) = \begin{cases} \dfrac{d}{1-d}, & (n-1)T_h \leq t \leq \dfrac{(2n-1)T_h}{2} \\[3mm] \dfrac{d}{1+d}, & \dfrac{(2n-1)T_h}{2} \leq t \leq n(T_h) \end{cases} \tag{3.18}$$

The voltage transfer ratio is shown to be discontinuous because of the unfolding bridge. Figure 3-9 shows the graphical representation of the voltage transfer ratio.

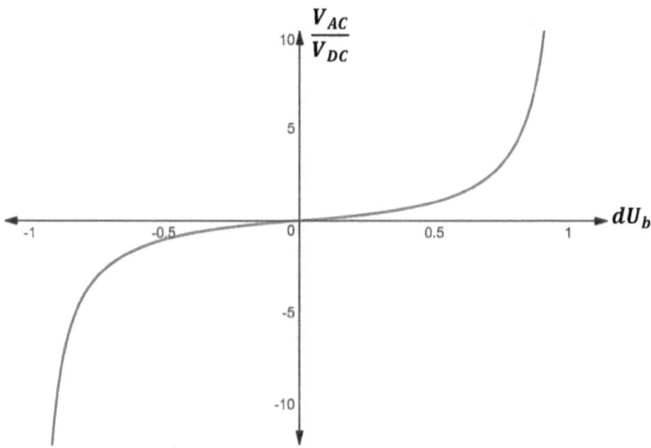


**Figure 3-9 Ideal voltage transfer as a function of the duty cycle**

The continuous-time duty cycle was multiplied by the unfolding function to show the transfer ratio as a continuous function. If all the DC power is converted into AC power i.e. efficiency is 100%, current transfer ratio is given by:

$$\frac{i_{AC}}{i_{DC}}(d) = \begin{cases} \dfrac{1-d}{d}, & (n-1)T_h \leq t \leq \dfrac{(2n-1)T_h}{2} \\[3mm] \dfrac{1+d}{d}, & \dfrac{(2n-1)T_h}{2} \leq t \leq n(T_h) \end{cases} \tag{3.19}$$

Equations (3.18) and (3.19) show how power is ideally transferred using the duty cycle as a time-averaging factor.

### *iii.    Output voltage generation*

Based on the current analysis of the inverter, the mathematical analysis was done to show how the output voltage is created. To achieve sinusoidal voltage, the duty cycle was made to vary, such that the gain achieved by the converter stage also varies. The duty cycle can be derived from the voltage transfer ratio in (3.18) as follows:

$$d = \begin{cases} \dfrac{V_m\sin(2\pi f_h t)}{V_m\sin(2\pi f_h t) + V_{DC}}, & (n-1)T_h \leq t \leq \dfrac{(2n-1)T_h}{2} \\[4mm] \dfrac{V_m\sin(2\pi f_h t)}{V_{DC} - V_m\sin(2\pi f_h t)}, & \dfrac{(2n-1)T_h}{2} \leq t \leq n(T_h) \end{cases} \tag{3.20}$$

Plotting the duty cycle expression shows the profile necessary to obtain the sinusoid. To make the duty cycle modulating signal continuous, it was multiplied by the unfolding function.

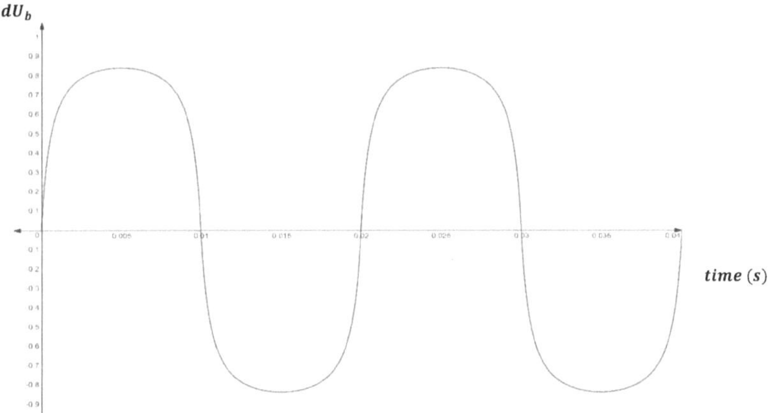


**Figure 3-10 DC-AC continuous time duty cycle modulating signal**

The waveform observed in Figure 3-10 shows the duty cycle starting from zero and varying steeply before slowing down to reach a peak duty cycle of 83%. This also shows that the inverter works both as a buck and boost converter to obtain the sinusoid. To obtain the output voltage, the product of the unfolding bridge function and the output of the Ćuk converter was taken i.e.

$$V_{AC} = V_{DC}\left(\frac{d}{1-d}\right)U_b \tag{3.21}$$

Evaluating the expression in (3.21), the desired output AC voltage was obtained:

$$V_{AC} = V_m\sin(2\pi f_h t) \tag{3.22}$$

Equation (3.22) represents the desired inverter output voltage that is being designed for. To illustrate this graphically the waveforms in Figure 3-11 were plotted.

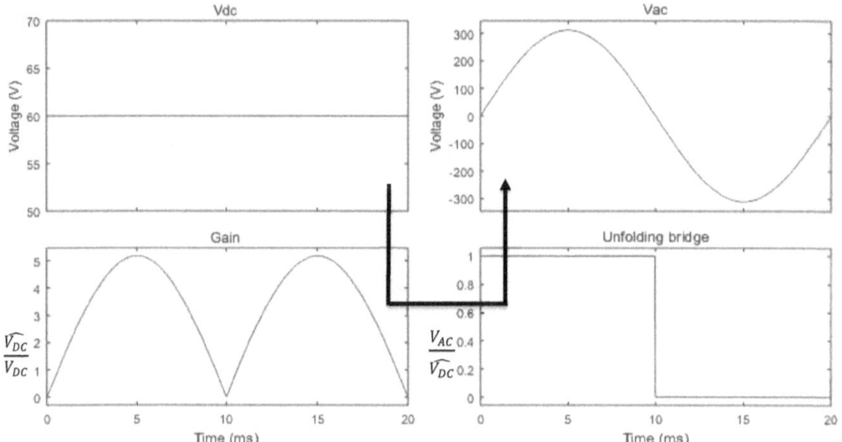



Figure 3-11 DC-AC output voltage generation

The first waveform represents the input DC voltage; the second waveform is the converter stage gain from the input DC voltage to the pseudo DC voltage; the third is the unfolding bridge function and last waveform is the actual output voltage. The DC input voltage is multiplied by the gain, to obtain the pseudo-DC voltage. The pseudo-DC is then multiplied by the unfolding function to obtain the required AC voltage.

*iv.* **The boundary between buck and boost regions**

In summary, the inverter changes from step up to step down depending on the output voltage desired and the input line voltage.

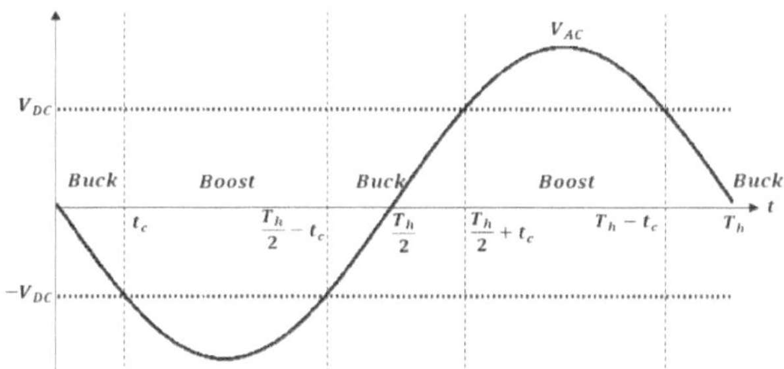


**Figure 3-12 Ćuk Inverter buck and boost mode transitions**

Based on Figure 3-12, the inverter changes from buck to boost mode four times in one period. The regions were described by looking at the first angle. At 50% duty cycle is the boundary between the buck and boost regions, it is also the point where the output AC voltage is equal to the input voltage; knowing this:

$$V_{DC} = V_{AC(50\% \text{ duty cycle})} \tag{3.23}$$

Substituting (3.23) into (3.22):

$$V_{DC} = V_m \sin(2\pi f_h t_c) \tag{3.24}$$

Where $t_c$ is the critical time when the inverter switches between buck and boost. Solving (3.24) for the critical time $t_c$ gives:

$$t_c = \frac{\sin^{-1}\left(\frac{V_{DC}}{V_m}\right)}{2\pi f_h} rad \tag{3.25}$$

This is the critical time when the input voltage equals the output voltage at 50% duty cycle and the first angle where boost mode starts operating. Expressing those regions mathematically:

$$Region \begin{cases} Buck, & \left(n\dfrac{T_h}{2} - t_c\right) \le t \le \left(n\dfrac{T_h}{2} + t_c\right) \\[3ex] Boost, & \left(n\dfrac{T_h}{2} + t_c\right) \le t \le (n+1)\dfrac{T_h}{2} - t_c \end{cases} \qquad (3.26)$$

where $n = 0,1,2,3 \dots$ is a whole number. The expression above shows how the inverter changes regions periodically to ensure a smooth sinusoidal waveform.

### 3.2.3   AC-DC power flow inverter equivalent circuit and operation

The bidirectional inverter operates as a charger during AC-DC power flow. During which, AC power is converter back to DC power. As with the DC-AC power flow, the equivalent circuit and operation were analysed. During this power flow equation (3.22) is the input voltage and the DC voltage is the output voltage. The equivalent circuit during AC-DC power flow is shown in Figure 3-13.

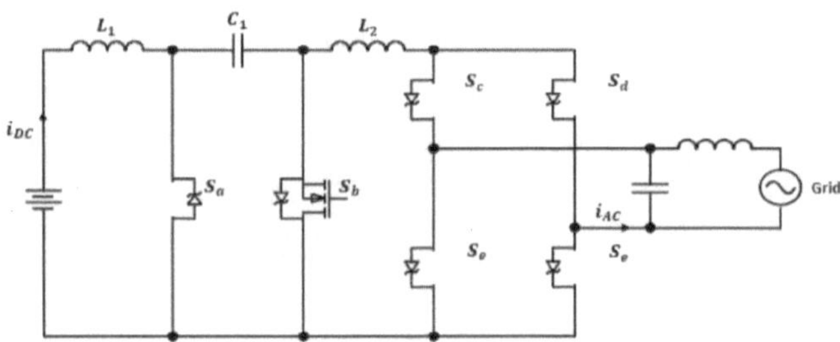


Figure 3-13 AC-DC power flow equivalent circuit

In this mode, only one switch is operational – the output switch $S_b$. All four Unfolding bridge switch body diodes and the body diode of switch $S_a$ are operational.

### i.        Current analysis

The power flows from right to left. Similar to the DC-AC power flow, the negative half cycle ON-state current analysis was done.

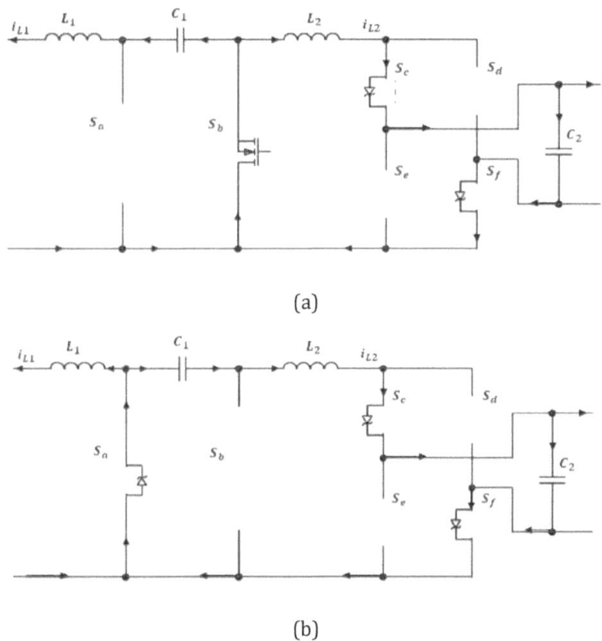

(a)

(b)

**Figure 3-14 AC-DC power flow negative half cycle high frequency ripple (a) ON-state (b) OFF-state**

During the ON-state as shown in Figure 3-14 (a), the output switch $S_b$ is operational. It charges the output inductor $L_2$ through body diodes from switches $S_c$ and $S_f$. The input capacitor $C_1$ is discharging through the input inductor $L_1$ and the load resistor. The internal diode of the input switch $S_a$ is reverse biased. The output capacitor is charging up. During the discharging state shown in Figure 3-14 (b), the body diodes from $S_c$ and $S_f$ are still operational. But the output switch $S_b$ is OFF. The output inductor $L_2$ then discharges through the diodes of the switches. The input capacitor $C_1$ charging through the body diode of the input switch $S_a$. The input inductor $L_1$ is discharging through the load and the body diode. This interchange between the charging and discharging states takes place for another half a period of the input voltage until the positive half cycle shown in Figure 3-15.

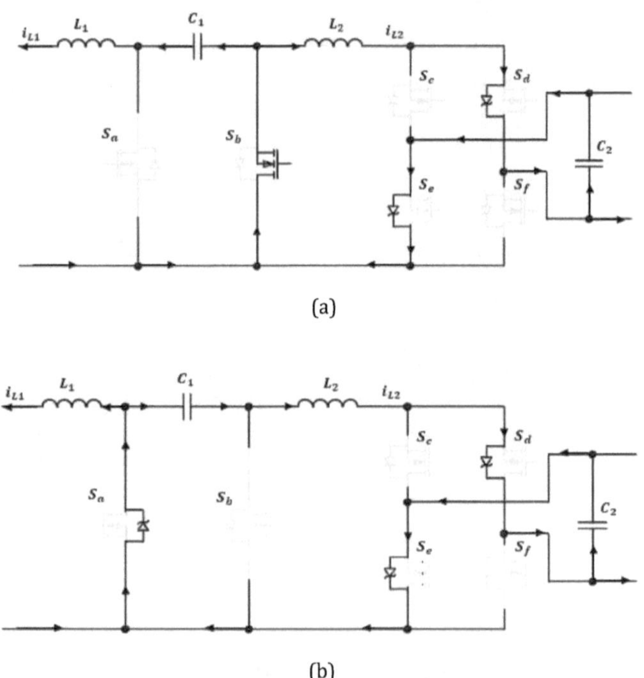


**Figure 3-15 AC-DC power flow positive half cycle high frequency ripple (a) ON-state (b) OFF-state**

The ON-state for the positive half cycle shown in Figure 3-15 (a) is the same as that of the negative half cycle. The only difference is that diodes $S_c$ and $S_f$ are reverse biased this time and diodes $S_d$ and $S_e$ are forward biased and conducting. The OFF-state in Figure 3-15 (b) is also the same as the negative. With the difference being in the diode conduction. After a full period, the negative half cycle starts again.

*ii.    Output voltage generation*

To achieve a DC voltage, transfer ratio of the inverter remains unchanged. Given the AC input voltage in (3.22). The unfolding bridge and rectified sine wave remain the same as the DC-AC power flow. The duty cycle required was derived as follows:

$$d = \begin{cases} \dfrac{V_{DC}}{V_m\sin(2\pi f_h t) + V_{DC}}, & (n-1)T_h \leq t \leq \dfrac{(2n-1)T_h}{2} \\[3ex] \dfrac{V_{DC}}{V_{DC} - V_m\sin(2\pi f_h t)}, & \dfrac{(2n-1)T_h}{2} \leq t \leq n(T_h) \end{cases} \tag{3.27}$$

The DC voltage to be achieved is $74V$ for charging a $60V$ battery as will be designed for in section 4.2. Plotting the duty cycle:

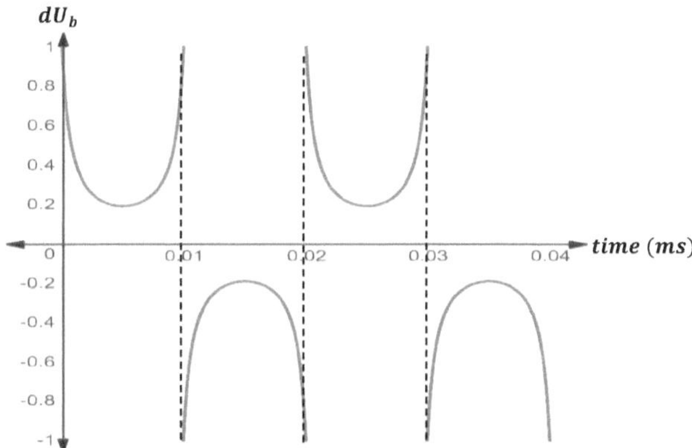



**Figure 3-16 AC-DC continuous time duty cycle modulating signal**

Figure 3-16 shows that the duty cycle is inverted compared to the inverter. It reaches a minimum duty value of 0.19 – this is the point where the peak amplitude of the input voltage is decreased to equal the desired DC voltage. The output voltage can then be obtained by:

$$V_{DC} = V_m\sin(2\pi f_h t)\,(U_b)\left(\frac{d}{1-d}\right) \tag{3.28}$$

This was again illustrated using waveforms as shown in Figure 3-17.

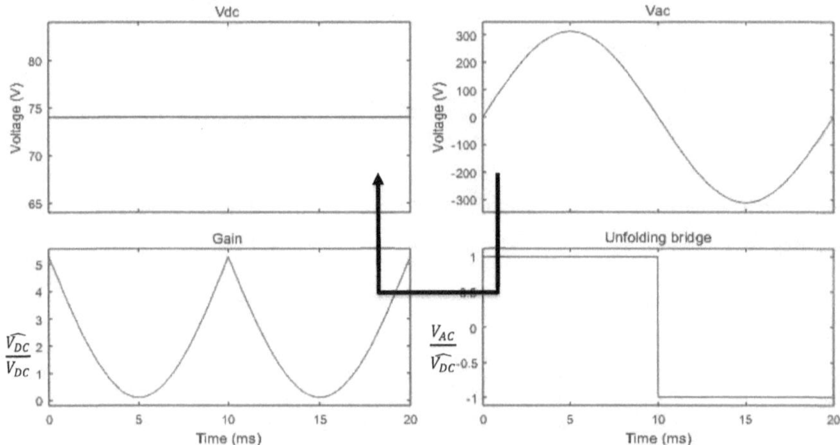

**Figure 3-17 AC-DC output voltage generation**

The input AC voltage is multiplied by the unfolding function to obtain the pseudo-DC voltage. This is then multiplied by the inverse gain to obtain the charging DC voltage.

### *iii.     Boundary between buck and boost regions*

The buck and boost regions of the reverse AC-DC power flow are like those of the DC-AC power flow.

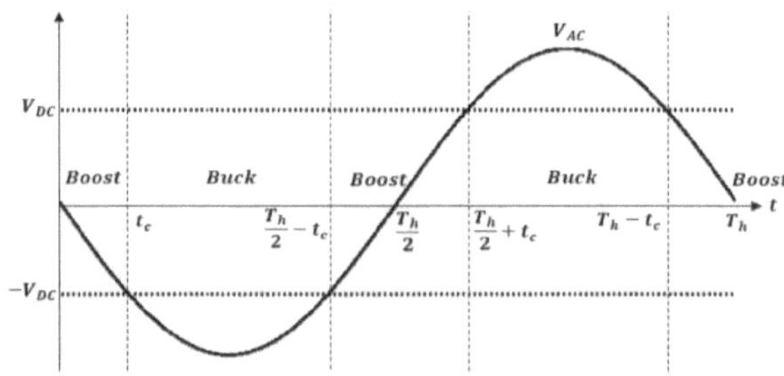


**Figure 3-18 AC-DC power flow region of operation transitions**

Figure 3-18 shows sections of the AC-DC power flow that alternate with respect to those in the DC-AC power flow of the inverter. The first angle is also the same as the DC-AC power flow and therefore, the regions though differ:

$$Region \begin{cases} Boost, & \left(n\dfrac{T_h}{2} - t_c\right) \leq t \leq \left(n\dfrac{T_h}{2} + t_c\right) \\[2em] Buck, & \left(n\dfrac{T_h}{2} + t_c\right) \leq t \leq (n+1)\dfrac{T_h}{2} - t_c \end{cases} \qquad (3.29)$$

where $n = 0,1,2,3 \ldots$ is a whole number. These regions also transition as shown to ensure that the varying magnitude of the output input AC voltage is converted into a constant DC output charging voltage.

## 3.3    Inverter passive components modes of operation

The bidirectional inverter was designed such that, during DC-AC power flow both inductor currents were continuous; this ensures that the efficiency of the inverter is high since any discontinuities in the inductor current increases the losses at the output. During AC-DC power flow, the inverter was designed such that, the coupling capacitor voltage is discontinuous; this ensures reactive power compensation and as such, the inverter operates as a power factor correction converter during this power flow [36].

### 3.3.1    *Continuous inductor current mode (CICM)*

In CICM, the inductor current does not reach zero for all time. It is always continuous. This mode is most the common mode which the front-end converter is used in. Due to it having less losses and while being easier to design than the other modes. Figure 3-19 shows the inductor current waveforms at the boundary between continuity and discontinuity of the inductor currents.

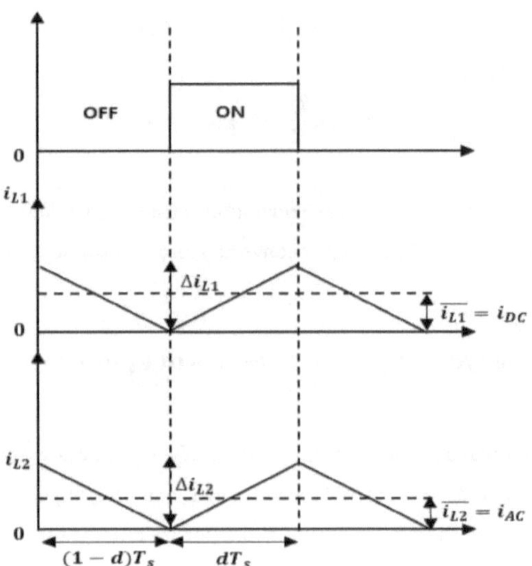


**Figure 3-19 Continuous conduction mode component waveforms**

Both input and output inductor currents are continuous but at the lowest current values before they become discontinuous. The average of the input inductor current $i_{L1}$ is the input current and is exactly half of the peak t peak ripple. The average of the output inductor $i_{L2}$ is the load current and is half the peak to peak ripple of the output inductor current. There are minimum values for the passive components required to keep the inverter operating in CICM; these were analysed.

### *i.    Minimum input inductor*

From the Figure 3-19, at the boundary between continuous and discontinuous mode, the average inductor current is given as:

$$\Delta i_{L1} = 2\overline{i_{L1}} \tag{3.30}$$

where $\overline{i_{L1}}$ is the average input inductor current [37]. Because of the capacitor charge balance, the input DC current is the average of the input inductor current. Knowing this fact, from the waveform above that:

$$\Delta i_{L1} = \frac{dV_{DC}}{f_s L_1} = 2 i_{DC} \qquad (3.31)$$

Since the input current is:

$$i_{DC} = i_{AC,peak} \frac{d}{1 - d} \qquad (3.32)$$

And the output current:

$$i_{AC,peak} = \frac{V_{AC,peak}}{R_L} \qquad (3.33)$$

Where $R_L$ is the AC load resistance. Therefore from (3.30),(3.31),(3.32) and (3.33):

$$\frac{dV_{DC}}{f_s L_1} = 2 \left(\frac{d}{1 - d}\right)^2 \left(\frac{V_{AC,peak}}{R_L}\right) \qquad (3.34)$$

Solving for the inductor:

$$L_{1min} = \frac{R_L (1 - d_{min})^2}{2 d_{min} f_s} \qquad (3.35)$$

Where $d_{min}$ is the duty cycle required to step down the peak grid voltage, to the battery charging voltage. From the equation (3.35), decreasing the value of the inductor brings the converter closer to discontinuity because the inductor current decreases. Decreasing the frequency also leads to discontinuity of the converter and the lower the load current, the closer the inverter gets to discontinuity.

*ii.*     ***Minimum output Inductor***

For the output inductor:

$$\Delta i_{L2} = \frac{2 V_{AC,peak}}{R_L} \qquad (3.36)$$

But because:

$$2i_{AC,peak} = \frac{dV_{DC}}{f_s L_{2min}} \tag{3.37}$$

From (3.36) and (3.37) it means that:

$$L_{2min} = \frac{R_L(1 - d_{min})}{2f_s} \tag{3.38}$$

The factor that affect the continuity of the output inductor current are the same as those of the input inductor current. Both these conditions must be met simultaneously to ensure continuous conduction mode.

*iii.* **Minimum coupling capacitor**

At the boundary between continuity and discontinuity of the coupling capacitor, the ripple coupling capacitor voltage is twice the output voltage i.e.

$$\Delta V_{c1} = 2V_{AC,peak} \tag{3.39}$$

Using the fact that:

$$\Delta V_{c1} = \frac{i_{L1}(1 - d_{min})}{f_s C_{1min}} \tag{3.40}$$

Therefore from  and :

$$2V_{AC,peak} = \frac{i_{L1}(1 - d_{min})}{f_s C_{1min}} \tag{3.41}$$

And simplifying the expression:

$$C_{1min} \geq \frac{d_{min}}{2f_s R_L} \tag{3.42}$$

This is the minimum coupling capacitor value needed to ensure continuity when the bi-directional inverter is converting DC power into AC power. Increasing the frequency brings the converter closer to discontinuity. Decreasing the load current also brings the inverter closer to discontinuity.

### 3.3.2    *Discontinuous capacitor voltage mode (DCVM)*

The inverter was designed such that, during AC-DC power flow, the coupling capacitor voltage is discontinuous. This mode is especially important for power factor correction applications. Although, the coupling capacitor voltage is discontinuous, both inductor currents are continuous – hence there are no inductor losses associated with the switch as would be in discontinuous inductor current mode (DICM). This mode also ensures soft switch-off which is desirable because it slows down the voltage rise in the switch, which can damage the switch. Another advantage is that it presents low switch current stress and relatively low ripple in the input current. The power factor obtained from this mode is almost unity [38].

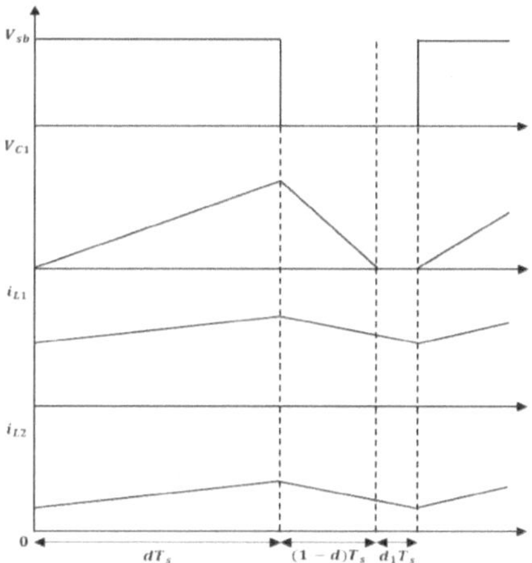


**Figure 3-20 Capacitor voltage waveform during on and off-states**

Figure 3-20 shows that when the switch $S_b$ is on the capacitor discharges for a fraction of the on-time and then completely discharges and remains at zero until the off-state where it starts charging up again. Unlike CICM, to ensure operation in this mode, these are maximum values instead of minimum values. The maximum coupling capacitor and inductor currents to ensure this mode is achieved were then determined.

*i.*        ***Maximum coupling capacitor***

The voltage appearing across the capacitor in this mode, from Figure 3-20 can be derived as follows [39].

$$V_{c1} = \begin{cases} \dfrac{i_{L2}T_s(1-d) - i_{L1}t}{C_1}, & 0 < t < d_1 T_s \\ 0, & d_1 T_s < t < d T_s \\ \dfrac{i_{L2}T_s(1-d)}{C_1}, & d T_s < t < T_s \end{cases} \tag{3.43}$$

The duty cycle $d_1$ is:

$$d_1 = \frac{i_{L2}}{i_{L1}}(1-d) \tag{3.44}$$

Since the input inductor current is:

$$i_{L2} = \frac{2 f_s C_1 V_{AC}}{(1-d)^2} \tag{3.45}$$

And the output inductor is given by:

$$i_{L1} = \frac{V_{DC}}{R_{Lb}} \tag{3.46}$$

Where $R_{Lb}$ is the equivalent battery DC Load impedance. From equations (3.43) to (3.46):

$$d_1 = \sqrt{2 f_s R_{Lb} C_1} \tag{3.47}$$

To operate in DCVM there needs to be a maximum input capacitor value to ensure that the capacitor time constant is smaller than the on-time of the converter i.e. $d_1 < d$; or in other terms

$$\sqrt{2 f_s R_{Lb} C_{1max}} < d \tag{3.48}$$

Therefore, from (3.48), the maximum capacitor value to ensure DCVM operation:

$$C_{1,max} < \frac{d_{max}^2}{2f_s R_{Lb}} \tag{3.49}$$

The equation (3.49) shows that this mode depends on how fast the device is being switched. The lower the frequency, the deeper the converter goes into DCVM operation. If the duty cycle is reduced the converter operates safely in DCVM. Low load resistances also ensure that the converter continues in DCVM. One of the main disadvantages of operating the Ćuk converter is this mode is that it increases the stress on the switch of the converter. The voltage level is the same as the maximum voltage appearing across the input capacitor i.e.

$$V_{stress} = \frac{i_{L1}}{C_1}(1-d)T_s \tag{3.50}$$

Evaluating the expression in (3.50):

$$V_{stress} = \frac{2}{(1-d_{max})}V_{DC} \tag{3.51}$$

The equation (3.51) shows that the higher the duty cycle, the higher the stress on the switch. To accommodate for this, a switch that can handle high voltages would need to be used.

*ii.*     ***Maximum input inductor***

The input inductor must be large enough to reduce the ripples arising from the AC voltage. From the grid side, the input impedance is given by:

$$r_{in} = \frac{V_{DC}}{i_{L2}} \tag{3.52}$$

From (3.45), this becomes:

$$r_{in} = \frac{(1-d)^2}{2f_s C_1 V_{AC}} \tag{3.53}$$

To reduce the ripples from the switching frequency the input impedance must be:

$$r_{in} \ll \frac{2\pi L_1}{T_h}$$
(3.54)

Therefore, substituting (3.53) into (3.54) and solving for the maximum input inductor.

$$L_{1,max} \ll \frac{1}{C_2 f_s^2}$$
(3.55)

*iii.*　　**Maximum output inductor**

Because the input impedance doesn't change and the condition for the input impedance also remains the same, the maximum output inductor becomes:

$$L_{2,max} \ll \frac{1}{C_2 f_s^2}$$
(3.56)

## 3.4　Bidirectional inverter performance assessment

The bidirectional inverter's performance assessment was analysed starting with DC-AC power performance and the AC-DC power flow performance.

### 3.4.1　DC-AC power flow performance review

To assess the performance of any inverter the following considerations must be considered.

*i.*　　**Output voltage**

The RMS voltage can be computed from:

$$V_{AC,RMS} = \sqrt{\frac{1}{\pi} \int_0^{\pi} V_m^2 \, sin^2(\omega t) \, d(\omega t)}$$
(3.57)

$$= \frac{V_m}{\sqrt{2}}$$

*ii.* **Load regulation**

The voltage variations must be as per power quality requirements. To determine the regulation, equation (3.58) is used.

$$\%load = \frac{|V_{omax} - V_{omin}|}{V_{onom}} \tag{3.58}$$

Where $V_{omax}$ is the output voltage at rated load current; $V_{omin}$ is the output voltage when the load current is minimum and $V_{onom}$ is the output voltage at the nominal load current.

*iii.* **Line regulation**

The regulation concerns variations in the input voltage.

$$\%line = \frac{|V_{vmax} - V_{vmin}|}{V_{vnom}} \tag{3.59}$$

Where $V_{vmax}$ is the output voltage when the input voltage is at its maximum; $V_{vmin}$ is the output voltage when the input voltage is at its minimum and $V_{vnom}$ is the output voltage when at the nominal input voltage.

*iv.* **Output voltage harmonic distortion**

The Fourier series analysis of any repeating waveform is given by:

$$f(t) = \frac{1}{2}a_0 + \sum_{h=1}^{\infty}(a_h \cos h\omega t + b_h \sin h\omega t) \tag{3.60}$$

where:

$$a_h = \int_0^{2\pi} f_h(t) \cos h\omega t \, d\omega t$$

$$b_h = \int_0^{2\pi} f_h(t) \sin h\omega t \, d\omega t$$

In terms of Fourier series decomposition RMS components $f_h(t)$:

$$F = \left( F_o^2 + \sum_{h=1}^{\infty} F_h^2 \right)^{\frac{1}{2}} \tag{3.61}$$

For an input voltage $V_{AC}(t) = V_m \sin \omega t$ its decomposition is:

$$V_{AC}(t) = \sqrt{2}V_1 \sin(\omega_1 t - \varphi_1) + \sum_{h=1}^{\infty} \sqrt{2}V_h \sin(\omega_h t - \varphi_h) \tag{3.62}$$

Where $V_1$ is the RMS fundamental voltage; $V_h$ is the RMS voltage from each harmonic. From equation (3.62), it's clear that if the fundamental is subtracted from a function, the harmonic spectrum of the function can be obtained, i.e.

$$V_{hAC,RMS} = V_{AC,RMS} - V_{1AC,RMS} \tag{3.63}$$

Therefore, the Total Harmonic Distortion is given by:

$$\%THD = 100 \times \frac{\sqrt{(V_{AC,RMS}^2 - V_{1,RMS}^2)}}{V_{1,RMS}} \tag{3.64}$$

v. ***Output power factor***

The average power of an electrical circuit is given by:

$$P = \frac{1}{T_s} \int_0^{T_s} V_{AC}(t) i_{AC}(t)\, dt \tag{3.65}$$

$$= V_{AC,RMS} i_{AC,RMS} \cos \varphi_1$$

Where $\varphi_1$ is the output PF angle. From (3.65) the Power factor is therefore given by:

$$PF = \frac{V_{1,RMS} i_{1,RMS}}{V_{AC,RMS} i_{AC,RMS}} \tag{3.66}$$

This output power factor indicates how much real power is delivered to the AC load output.


### 3.4.2    AC-DC power flow Performance review


Bidirectional inverters operate as normal inverters as they convert DC power into AC power. During AC-DC power conversion however, they operate as PWM line frequency rectifies. Their performance was analysed.


#### i.        *Effect of the AC source inductance on the DC output voltage*


Since the bidirectional inverter operates as PWM rectifier during reverse AC-DC power flow, it is affected similarly to a rectifier. Figure 3-1 shows that the AC source has a source inductance – as with any AC source – as such, the output voltage of the unfolding bridge body diodes will be:

$$\widehat{V_{DC}} = \frac{1}{\pi}\left( \int_{0}^{\pi} V_m sin\omega t \, d(\omega t) - \int_{0}^{\delta} V_m sin\omega t \, d(\omega t) \right)$$

(3.67)

Where $\delta$ is the dead band interval. Evaluating the integral:

$$\widehat{V_{DC}} = \frac{2V_m}{\pi} - \frac{1}{\pi}V_m(1 - \cos(\delta))$$

(3.68)


To analyse the effects of this inductance, a constant DC load current is assumed. All diodes have a small turn-on voltage, assuming a fast recovery time of the diode, the transition from OFF-ON is instantaneous. This turn-on voltage results in dead bands in the rectified voltage at zero crossings. But when a source inductance is present, it prevents the current from instantaneously rising to the load current instead, the current rises gradually as shown in Figure 3-21. This increases the dead bands of the rectified voltage, thereby lowering the output voltage.

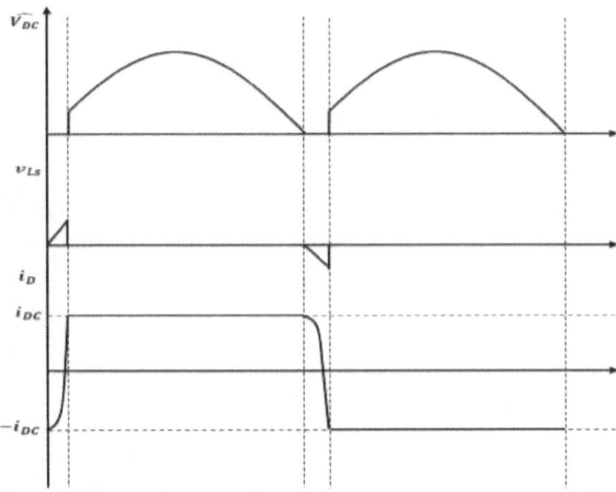

Figure 3-21 Effect of source inductance on the unfolding bridge output

To solve for the term $\cos(\delta)$, the dead band intervals were evaluated. During the dead bands, the voltage induced across the source inductance is given by:

$$v_{Ls} = L_s\frac{di_s}{dt} = V_m sin\omega t$$

(3.69)

Where, $V_{max}$ is the grid peak voltage and $\omega$ is the grid fundamental frequency. Integrating both sides of the equation from 0 to until max load current is reached across the diode, gives:

$$\int_0^\delta V_m sin\omega t\, d(\omega t) = \omega L_s \int_0^{i_{DC}} di_s$$

(3.70)

Where $i_{DC}$ is the maximum load current. Evaluating the integral:

$$V_m(1 - \cos(\delta)) = \omega L_s i_{DC}$$

(3.71)

Simplifying the expression:

$$\cos(\delta) = 1 - \frac{\omega L_s i_{DC}}{V_m} \tag{3.72}$$

Substituting (3.72) into (3.68):

$$\widehat{V_{DC}} = \frac{2V_m}{\pi}\left(1 - \frac{\omega L_s i_{DC}}{V_m}\right) \tag{3.73}$$

From equation (3.28), the DC output voltage is then:

$$V_{DC}(i_{DC}) = \frac{2d_{min}V_m}{(1 - d_{min})\pi}\left(1 - \frac{\omega L_s i_{DC}}{V_m}\right) \tag{3.74}$$

Equation (3.74) shows that if the inverter is operating at full load, the output voltage will be at its lowest and if the grid voltage fluctuates, the output DC voltage will be at its lowest. Figure 3-22 shows graphically the DC output voltage as a function of the load current. The load current was swept from zero to a maximum battery charging current of $12A$. The peak AC voltage was kept constant at $\sqrt{2}220V$. The AC source inductance was sized at $5mH$. The minimum duty cycle was kept at 19.20% as required to step down the peak AC voltage to $74V$, to charge a $60V$ battery.

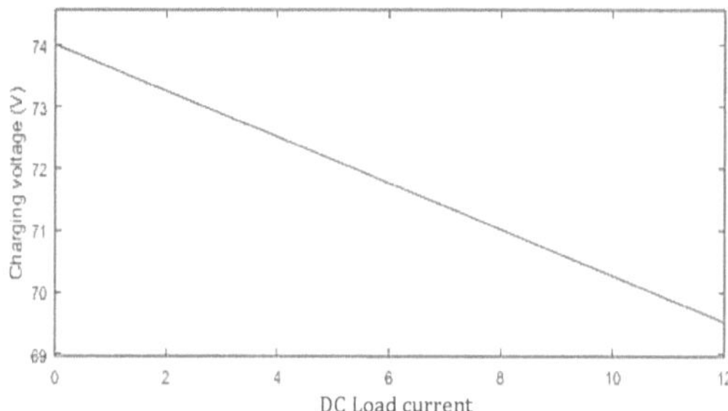


**Figure 3-22 Effect of source inductance on the output DC charging voltage**

The relationship between the charging DC voltage and load current under the influence of the AC source inductance is linear. The DC voltage is maximum when the load current is zero and minimum when the load current is maximum. This is because the more load strain is exerted on the system, the higher the current drawn from the AC source which increases the voltage drop across the source inductor.

*ii.* ***Unfolding bridge output efficiency***

The efficiency at the output of the unfolding bridge is given by:

$$e_{ff} = \frac{i_{DC}^2 R_c}{i_{AC,RMS}^2 \left(R_f + R_c\right)}$$

(3.75)

Where $R_c$ is the Ćuk converter output impedance seen by the diode bridge; $R_f$ is the diode forward resistance. But the DC current is equivalent to:

$$i_{DC} = \frac{2i_m}{\pi}$$

(3.76)

And:

$$i_{AC} = \frac{i_m}{\sqrt{2}}$$

(3.77)

Therefore, from (3.75), (3.76) and (3.77):

$$e_{ff} = \frac{2\sqrt{2}}{\pi}\left(\frac{R_c}{R_f + R_c}\right)$$

(3.78)

Therefore, during AC-DC power flow the efficiency will be maximum at 81.2%. The total efficiency will be product of the diode bridge efficiency and the converter stage efficiency since they are cascaded. This will be shown in detail in section 3.7.4ii.

*iii.* ***Line current total harmonic distortion***

With any rectifier, there is a distortion in the line current drawn from the AC supply. The line voltage does not exhibit this phenomenon. The line current waveform can be expressed as follows:

$$i_{AC,RMS} = i_{1,RMS} + \sum_{h=2i+1}^{\infty} i_{h,RMS}$$

(3.79)

Where $i_{1,RMS}$ is the fundamental frequency RMS value; $i_{h,RMS}$ the odd harmonics RMS of the current; $h$ the odd harmonic number; $i = 1,2,3\ldots\ldots$

This then evaluates to:

$$i_{AC,RMS} = \sqrt{2}i_{1,RMS}\sin(\omega_1 t) + \sum_{h=2i+1}^{\infty} \sqrt{2}i_{h,RMS}\sin(\omega_h t) \tag{3.80}$$

The expression above shows that if the drawn current has no harmonics, it will only have the fundamental, which is purely sinusoidal.

$$THD_\% = \frac{100\sqrt{i_{AC,RMS}^2 - i_{1,RMS}^2}}{i_{1,RMS}} \tag{3.81}$$

For practical implementation the THD should be less than 8% as per power quality requirements.

*iv.*     ***Input power factor***

The distortion in the input current causes the power factor to drop from unity. This leads to traditional rectifiers having a poor power factor. The power factor can be calculated by considering the Displacement Power Factor (DPF) which is the phase difference between the supply voltage and fundamental frequency of the line current.

$$PF = DPF\frac{i_{1,RMS}}{i_{AC,RMS}} \tag{3.82}$$

If the phase difference between the supply voltage and fundamental current waveform is 0, The DPF is 1 since the cosine of zero is one.

## 3.5     Effects of switching dead-time

The Ćuk inverter uses 6 switches: 2 in the converter stage and 4 in the unfolding stage. As such, an analysis of the dead-time inserted in the switches and its effects were analysed.

### 3.5.1    *Signal analysis*

Considering the inverter shown in Figure 3-1, switches $S_a$ and $S_b$ during the converter stage, are conjugates – when one is ON the is OFF and vice versa; and $S_d, S_e$ and $S_c, S_f$ during the unfolding stage, are also conjugates, as such – there is an instant in time where the switch signals overlap, and they are both in their ON state.

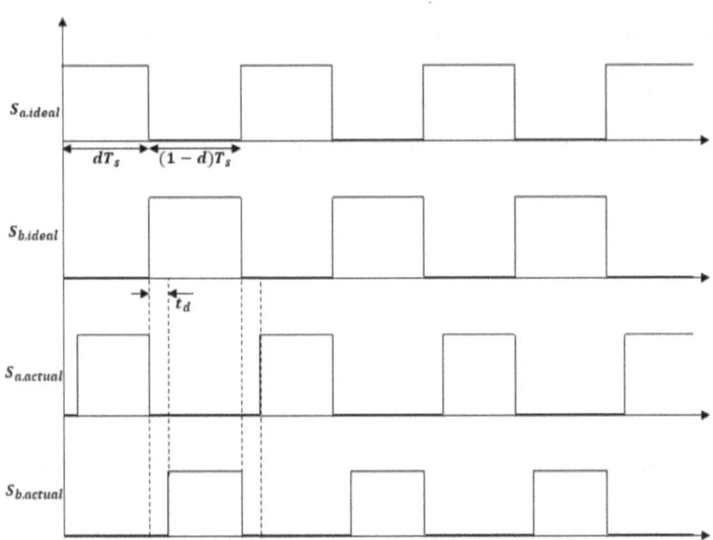



**Figure 3-23 Ćuk Inverter converter stage switches dead-time insertion**

Figure 3-23 shows how conjugate signals switch. As the first signal goes from ON to OFF, the second signal switches from OFF to ON and they overlap. The current at the drain of the switch will have a large spike during switch OFF – known as shoot-through – which can damage switches. This occurs because of the switches being in their OFF-state at the same instant momentarily. If this is compensated for, the short circuiting during this time, can damage the switches due to the large collector current. To prevent this, dead time is introduced in the switching signals. It is a time delay, $t_d$ in the signals that ensures that there is no overlap between complementing signals. Although it prevents shoot through, it presents some issues to the design that will be analysed. Since during DC-AC power flow all six switches are used, the analysis focused on the effects of dead time on the AC output voltage.

### 3.5.2 Converter stage dead-time effects on the AC output voltage

The signals in Figure 3-24 show the input and output inductor voltages with dead time inserted.

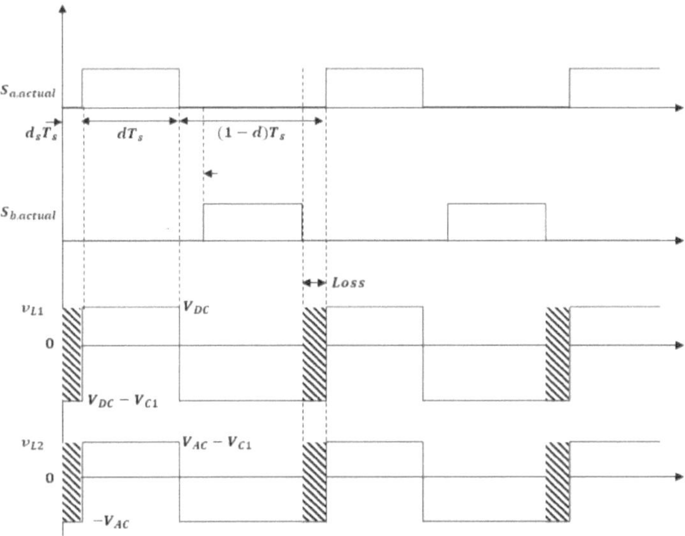



Figure 3-24 Inductor voltages with dead-time inserted

The inductor voltages show losses associated with the dead time. The effects on the RMS output voltage and its %THD are analysed.

#### i. Effects on the RMS output voltage

To analyse the effects of this dead time on the output AC voltage, equation (3.1) was considered. From the equation, the RMS output voltage can be obtained using equation (3.57) as:

$$V_{AC,RMS} = -\frac{V_{DC}}{\sqrt{2}}\left(\frac{d_{max}}{1 - d_{max}}\right) \tag{3.83}$$

The dead time $t_d$ shown in Figure 3-24 contributed to a loss in duty cycle in both the ON and OFF states. This duty cycle loss is given by:

$$d_s = \frac{t_d}{T_s}$$ (3.84)

The inverter operates in CICM and using the positive instantaneous inductor current mode as will be operated, the average inductor voltage is given by zero therefore, from Figure 3-24, the inductor voltage intervals can be described as follows:

$$\int_0^{d_s T_s} 0 \, dt + \int_{d_s T_s}^{d T_s} V_{DC} \, dt + \int_{d T_s}^{T_s} (V_{DC} - V_{c1}) \, dt = 0$$ (3.85)

Evaluating the summation integrals:

$$(d - d_s) V_{DC} T_s + (V_{DC} - V_{c1})(1 - d) T_s = 0$$ (3.86)

Simplifying the expression (3.86) the capacitor voltage:

$$V_{c1} = V_{DC} \frac{d_s - 1}{d - 1}$$ (3.87)

For the output inductor voltage $v_{L2}$:

$$\int_0^{d_s T_s} 0 \, dt + \int_{d_s T_s}^{d T_s} (\widehat{V_{DC}} - V_{c1}) \, dt + \int_{d T_s}^{T_s} \widehat{V_{DC}} \, dt = 0$$ (3.88)

Evaluating the integral in (3.88):

$$(d - d_s)(V_{c1} - \widehat{V_{DC}}) T_s + \widehat{V_{DC}} (1 - d) T_s = 0$$ (3.89)

Simplifying the expression above:

$$V_{c1} = \widehat{V_{DC}} \frac{d - d_s}{d - 1}$$ (3.90)

From the coupling capacitor voltages in (3.87) and (3.90), the voltage transfer ratio can be obtained to be:

$$\frac{\widehat{V_{DC}}}{V_{DC}} = \frac{d - d_s}{1 - d} \tag{3.91}$$

The RMS output voltage as a function of the duty cycle dead time, using (3.57) can therefore be obtained as:

$$V_{AC,RMS}(d_s) = \frac{V_{DC}}{\sqrt{2}} \left( \frac{d_{max} - d_s}{1 - d_{max}} \right) \tag{3.92}$$

The equation (3.92) shows how the dead time affects the peak amplitude of the output voltage. The power quality for the inverter as specified using the EN 501060 standard, requires the RMS magnitude to be $\pm 10\%$ of $220V$ – which is $198V$ or higher in this case. As such, the graph in Figure 3-25 shows the RMS output voltage as a function of the dead time within the voltage regulation range.

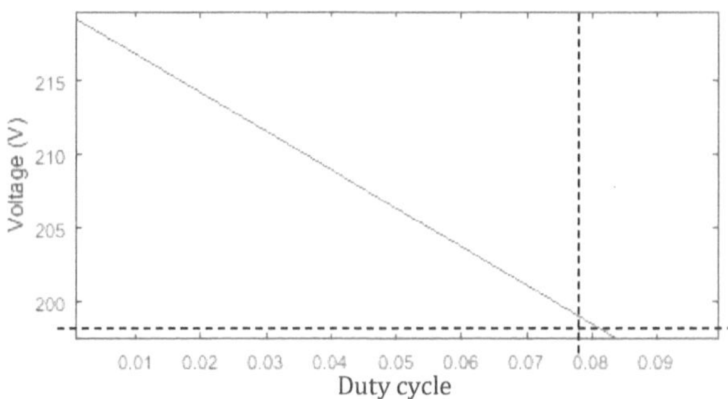


**Figure 3-25 RMS output voltage as a function of the dead time duty cycle**

The relationship between the voltage and dead time duty is linear in this region. As the dead time inserted into switches increases, the RMS output voltage decreases linearly until it falls below the regulation range. The dead time duty cycle where the voltage falls below the regulation range is 0.079, which – using equation (3.84) – is a dead time of $3.95\mu s$. Therefore, theoretically the dead time chosen for the converter stage switches must be less than this.

The harmonic distortion at the output voltage, as described by equation (3.64), is considered as the dead time varies from 0 until the %THD rises above the requirement. Substituting equation (3.92) into equation (3.64)gives:

$$\%THD(d_s) = 100 \times \frac{\sqrt{\left(\frac{V_{DC}}{\sqrt{2}}\left(\frac{d_{max}-d_s}{1-d_{max}}\right)^2 - \left(\frac{V_m}{\sqrt{2}}\right)^2\right)}}{\frac{V_m}{\sqrt{2}}} \tag{3.93}$$

The equation above shows the percentage THD as a function of the dead time duty cycle. The DC voltage was kept at $60V$ and the peak AC voltage $V_m$ was kept at $311.12V$ and lastly the maximum duty cycle was kept at 0.84. The graphical analysis in Figure 3-26, shows the %THD of the output voltage within the range of the EN 501060 standard of $< 8\%$.

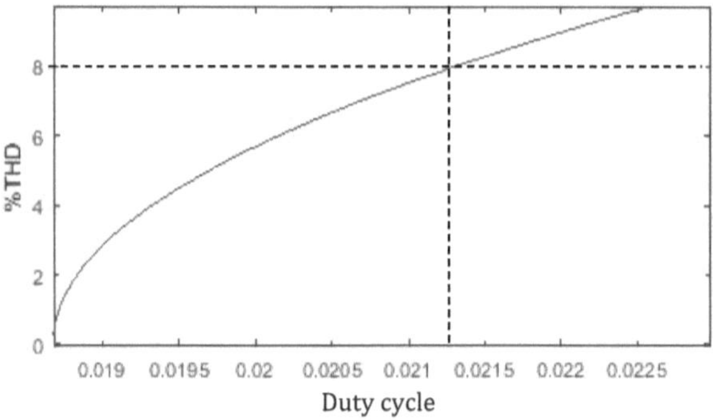

Figure 3-26 %THD as a function of the dead time duty cycle

The %THD initially rises steeply with a non-linear relationship as the dead time duty cycle increases. It then increases linearly until it is beyond the 8% mark at a dead time duty cycle of 0.0213 which – using equation (3.84) is a dead time of $1.065\mu s$. This dead time is less than that of the RMS requirement meaning that the dead time affects the power quality requirement of the THD more.

### 3.5.3    Unfolding stage dead time effects on the AC output voltage

The effects of dead time in the unfolding stage was assessed. When a sinusoidal signal passes through the bridge, it also appears with dead time as shown.

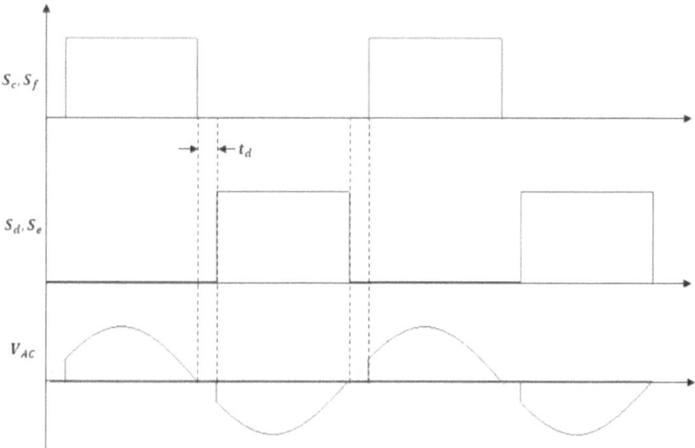


**Figure 3-27 Unfolding bridge dead time effects on output voltage of Ćuk inverter**

The sinusoidal output voltage shown in Figure 3-27 had a dead band of the same time as the dead time. This output voltage is obviously no longer purely sinusoidal, which in turn affects the harmonics of the output voltage. It also lowers the RMS value of the output voltage which in turn lowers the power at the output. These were discussed in detail.

### i.    *Effects on the RMS output value*

The RMS voltage at the output decreases. To analyse the RMS changes, a single AC voltage period from Figure 3-27 was analysed.

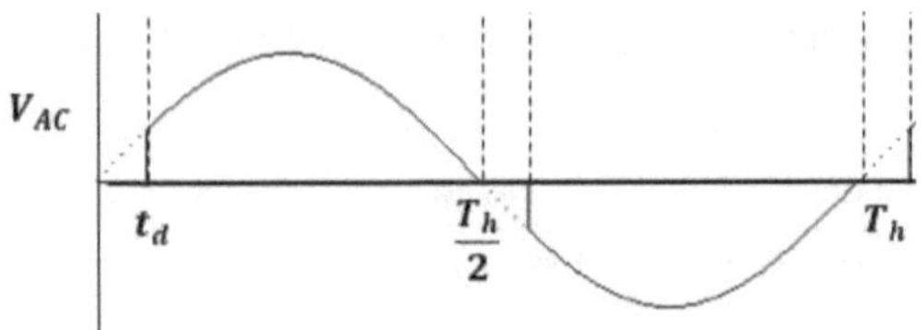

**Figure 3-28 Single output voltage period showing dead time effects**

For a given sinusoidal voltage shown in Figure 3-28 where $T_h$ is the grid fundamental period, the RMS of the voltage, from (3.57) and by using the trigonometric identity:

$$sin^2(\omega t) = \frac{1 - cos(2\omega t)}{2} \tag{3.94}$$

And substituting (3.94) into (3.57), using the intervals from Figure 3-31, the equation becomes:

$$V_{AC,RMS} = \sqrt{\frac{V_m^2}{T_h}\left(\int_0^{t_d} 0 \, d(\omega t) + \int_{t_d}^{\frac{T_h}{2}} \frac{1 - cos(2\omega t)}{2} d(\omega t)\right)} \tag{3.95}$$

Evaluating the integral in (3.95):

$$V_{AC,RMS} = \sqrt{\frac{V_m^2}{T_h}\left(\int_{t_d}^{\frac{T_h}{2}} \frac{1}{2} d(\omega t) - \int_{t_d}^{\frac{T_h}{2}} \frac{cos(2\omega t)}{2} d(\omega t)\right)} \tag{3.96}$$

Finally, the RMS voltage as a function of dead time becomes:

$$V_{AC,RMS}(t_d) = \frac{V_m}{\sqrt{2T_h}}\sqrt{\left(\frac{T_h}{2} - t_d\right) + \frac{sin(2t_d)}{2}} \tag{3.97}$$

The equation (3.97) above shows the RMS output voltage as a function of dead time. When the dead time is zero, the voltage RMS reduces to the ideal expression i.e. Substituting $t_d = 0$ into (3.97) gives:

$$V_{AC,RMS}(0) = \frac{V_m}{\sqrt{2T_h}} \sqrt{\left(\frac{T_h}{2} - 0\right) + \frac{\sin(0)}{2}}$$

$$V_{AC,RMS}(0) = \frac{V_m}{\sqrt{2}}$$

To illustrate the effect of dead time on the RMS, the function was plotted for a sinusoid of an amplitude of $311.23V$. The range of dead time was chosen such that the output voltage lies within the required voltage regulation range.

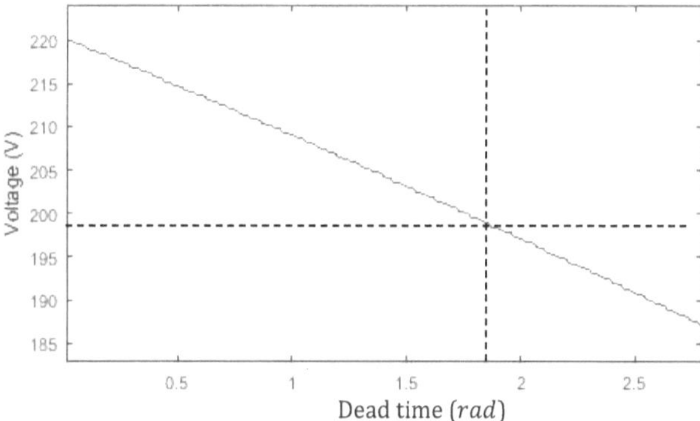



**Figure 3-29 RMS output voltage as function of dead time**

The graph in Figure 3-29 shows that the RMS voltage decreases with increasing dead time. When the dead time is zero the RMS is maximum 220V. When the dead time is half a period at $\pi$ radians (which is $180^o$), the RMS voltage is zero i.e. when the dead time is maximum the output voltage is would be zero. This is important when designing as the dead should be as small as possible but also enough to ensure that switching signals do not overlap. When the output voltage falls below the regulation range the dead time is $1.52ms$. This dead time is much higher than the dead time required during the converter stage i.e. it would take a larger dead time value during the unfolding stage for the output voltage to fall below the regulation range. Theoretically, the dead time chosen and inserted into the unfolding stage section should not affect the RMS voltage significantly, since the dead time inserted into switches is normally below $5us$.

*ii.*      ***Effects on the total harmonic distortion***

The dead time at the unfolding stage will also have a significant effect on the THD at the output. To analyse this effect, equation (3.64) was considered. Taking equation (3.97) and substituting it into equation (3.64) gives:

$$\%THD(t_d) = 100 \times \frac{\sqrt{\left(\left(\frac{V_m}{\sqrt{2T_h}}\sqrt{\left(\frac{T_h}{2}-t_d\right)+\frac{\sin(2t_d)}{2}}\right)^2 - \left(\frac{V_m}{\sqrt{2}}\right)^2\right)}}{\frac{V_m}{\sqrt{2}}} \tag{3.98}$$

Equation (3.70) shows the %THD as a function of dead time. Plotting the equation within the required THD range for graphical analysis, yielded the following:

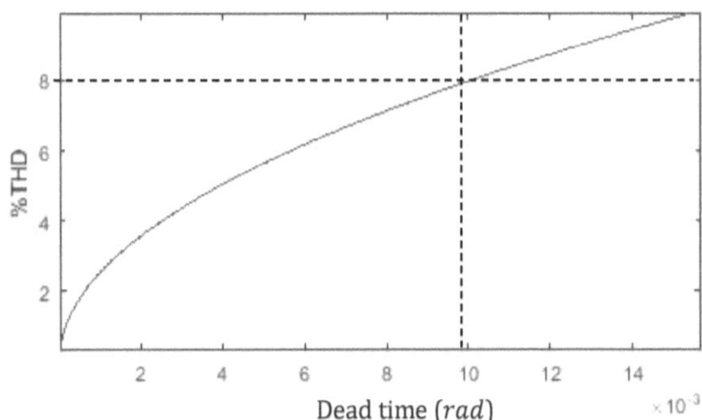


Figure 3-30 Dead time effects on the %THD at the output voltage

Figure 3-30 shows that the relationship between %THD and the dead time non-linear. The %THD rises steeply from zero before showing a linear region. The dead time when the %THD rises above the required 8%, is 63.66$\mu s$. Therefore, the dead time required to meet the %THD requirement is much lower than the dead time required to meet the voltage regulation requirement – this was also the case with the converter stage dead time. This is because of the steep rise in THD during the non-linear region. The dead time inserted into the unfolding stage also should not affect the THD significantly.

The theoretical maximum dead time required to meet the voltage RMS and THD power quality requirements from both the converter stage and unfolding bridge was compared in Table 3-1.

**Table 3-1 Maximum dead time during converter stage and unfolding bridge required to meet the output voltage and THD power quality requirements**

|  | Output voltage dead time ($\mu s$) | THD dead time ($\mu s$) |
|---|---|---|
| **Converter Stage** | 3.95 | 1.07 |
| **Unfolding Bridge** | 1520 | 63.66 |

The converter stage maximum dead time is shown to be much lower than that of the unfolding bridge because of the much lower frequency of the unfolding bridge. The THD maximum is shown to be lower than the voltage maximum dead time during both converter stage and unfolding stage. This is because of the non-linear relationship that the THD has with dead time compared to the linear relationship that the output voltage has with the dead time.

## 3.6     Parasitic effect on the inverter's power transfer

The parasitic elements have been shown to impact the power transfer of power converters. They limit the transfer ratio depending on how many parasitic elements there are in the circuit. The Ćuk inverter has 3 parasitic elements sources: 2 inductors, 2 capacitors and 6 switches. Studies have shown that the parasitic resistance in an inductor is the dominant parasitic element over the capacitor and switch parasitic resistances. Therefore, only the inductor's parasitic effects will be considered.

The following assumptions were made in the analysis:

- Capacitor and switches ESR's are negligible

- The capacitor currents average current is zero

- The coupling capacitor was large enough such that the voltage appearing it was constant

### 3.6.1    *Mathematical analysis*

The parasitic equivalent circuit for analysis is shown in Figure 3-31.

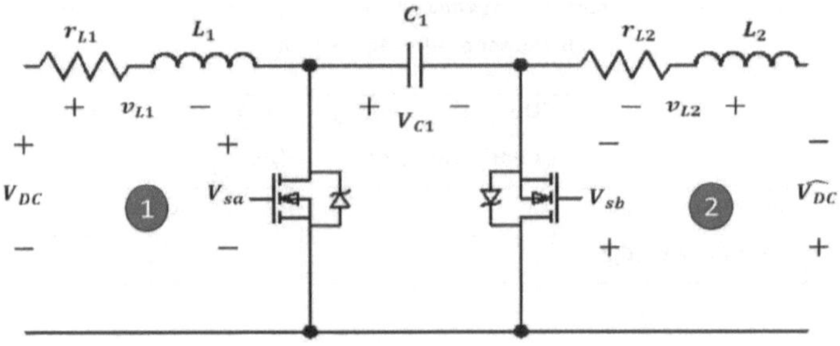

Figure 3-31 Ćuk converter equivalent circuit for parasitic gain analysis

Using the duty cycle as a time averaging factor, a clockwise loop for loop 1 and an anti-clockwise loop for loop 2, the voltage appearing across the DC and pseudo DC voltages can be defined as follows:

$$V_{DC} = \overline{v_{L1}} + \overline{V_{sa}} \qquad (3.99)$$

For the DC voltage and:

$$\widehat{V_{DC}} = \overline{V_{sb}} - \overline{v_{L2}} \qquad (3.100)$$

For the pseudo DC voltage. The average voltage appearing across the switch $S_a$:

$$\overline{V_{sa}} = \frac{1}{T_s}\left( \int_0^{dT_s} (V_{DC} - V_{DC})dt + \int_{dT_s}^{T_s} (V_{DC} - (V_{DC} - V_{c1})dt \right) \qquad (3.101)$$

Evaluating the integral gives:

$$\overline{V_{sa}} = (V_{DC} - V_{DC})d + (1-d)V_{c1} \qquad (3.102)$$

Therefore from (3.102) the voltage appearing across switch $S_a$ is given by (3.103) :

$$\overline{V_{sa}} = (1-d)V_{c1} \qquad (3.103)$$

Solving for the average voltage appearing across the switch $S_b$:

$$V_{sb} = \frac{1}{T_s}\left(\int_0^{dT_s} (\widehat{V_{DC}} - (V_{c1} - \widehat{V_{DC}})dt + \int_{dT_s}^{T_s} (\widehat{V_{DC}} - \widehat{V_{DC}})dt\right) \qquad (3.104)$$

Evaluating the integral in (3.104) gives:

$$\overline{V_{sb}} = dV_{c1} + (\widehat{V_{DC}} - \widehat{V_{DC}})(1 - d) \qquad (3.105)$$

Therefore from (3.105), the voltage appearing across switch $S_b$ is given by (3.106):

$$\overline{V_{sb}} = dV_{c1} \qquad (3.106)$$

The average inductor voltage for the input inductor is given by (3.107):

$$\overline{v_{L1}} = \overline{i_{L1}} r_{L1} \qquad (3.107)$$

Where $\overline{i_{L1}}$ the average input inductor voltage. From the coupling capacitor current:

$$(1 - d)\overline{i_{L1}} = di_{AC} \qquad (3.108)$$

Solving (3.108) for the input inductor current:

$$\overline{i_{L1}} = -d\frac{i_{AC}}{(1 - d)} \qquad (3.109)$$

Substituting for the output current in (3.109) the equation becomes:

$$\overline{i_{L1}} = -d\frac{\widehat{V_{DC}}}{R_L(1 - d)} \qquad (3.110)$$

And substituting (3.110) into (3.107) the average input inductor voltage is given by (3.111):

$$\overline{v_{L1}} = -d\,\frac{\widehat{V_{DC}}\,r_{L1}}{R_L(1-d)} \tag{3.111}$$

The average output inductor voltage is given by:

$$\overline{v_{L2}} = \overline{i_{L2}}\,r_{L2} \tag{3.112}$$

Where $\overline{i_{l2}}$ the average output inductor voltage and is equal to the output current, therefore (3.112) becomes:

$$\overline{v_{L2}} = \frac{\widehat{V_{DC}}\,r_{L2}}{R_L} \tag{3.113}$$

Using the input and output voltage equations and substituting the parameters i.e. substituting the equations (3.103) and (3.111) into (3.99) the input voltage becomes:

$$V_{DC} = -d\,\frac{\widehat{V_{DC}}\,r_{L1}}{R_L(1-d)} + (1-d)V_{c1} \tag{3.114}$$

Substituting (3.106) and (3.113) into (3.100) the output voltage becomes:

$$\widehat{V_{DC}} = dV_{c1} - \frac{\widehat{V_{DC}}\,r_{L2}}{R_L} \tag{3.115}$$

Solving (3.114) and (3.115) for the coupling capacitor voltage $V_{c1}$ and equating the equations gives:

$$\frac{V_{DC} + d\,\dfrac{\widehat{V_{DC}}\,r_{L1}}{R_L(1-d)}}{(1-d)} = \frac{\widehat{V_{DC}} + \dfrac{\widehat{V_{DC}}\,r_{L2}}{R_L}}{d} \tag{3.116}$$

Solving the equation (3.116) for the converter stage voltage transfer ratio as a function of the duty cycle:

$$\frac{\widehat{V_{DC}}}{V_{DC}}(d) = \frac{d}{\frac{(1-d)r_{L2}}{R_L} + \frac{d^2 r_{L1}}{(1-d)R_L} + (1-d)} \tag{3.117}$$

The inverter voltage transfer ratio after the unfolding stage is then given by:

$$\frac{V_{AC}}{V_{DC}}(d) = \begin{cases} \dfrac{d}{\dfrac{(1-d)r_{L2}}{R_L} + \dfrac{d^2 r_{L1}}{(1-d)R_L} + (1-d)}, & V_{AC} > 0 \\[4ex] \dfrac{d}{\dfrac{(1+d)r_{L2}}{R_L} + \dfrac{d^2 r_{L1}}{(1+d)R_L} + (1+d)}, & V_{AC} \leq 0 \end{cases} \tag{3.118}$$

If the ESRs for both inductors ($r_{L2}$ and $r_{L1}$) are 0, then the equation (3.117) becomes:

$$\frac{\widehat{V_{DC}}}{V_{DC}}(d) = -\frac{d}{\frac{(1-d)(0)}{R_L} + \frac{d^2(0)}{(1-d)R_L} + (1-d)}$$

Which gives:

$$\frac{\widehat{V_{DC}}}{V_{DC}}(d) = -\frac{d}{(1-d)}$$

Which gives the ideal converter stage voltage transfer ratio which shows that the gain depends on the ratio between the ESR resistances and the Load Resistance. It also shows that input inductor ESR $r_{L1}$ affects the gain more than the output inductor ESR $r_{L2}$ this is because the input current is generally larger than the output current.

### 3.6.2    *Graphical analysis*

To visualise the effects of the ESR resistance on the gain, the voltage transfer ratio in (3.90) was plotted against the duty cycle for various ratios of load-ESR.

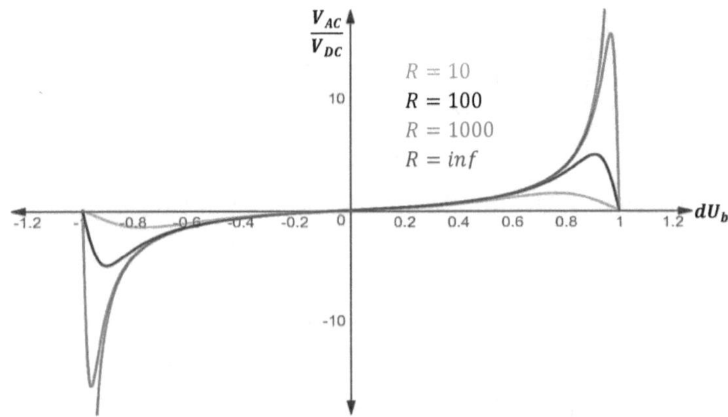


**Figure 3-32 Voltage transfer ratio at various load-ESR ratios**

From Figure 3-32, the load-ESR ratio is given by (3.119):

$$R = \frac{R_L}{(r_{L1} + r_{L2})} \tag{3.119}$$

The curve shown in blue shows the ideal case with no ESR resistances; this was also shown in Figure 3-9. It is shown to be infinite and does not converge like the other curve. The green curve is when the ratio is 10 and it peaks at $\pm2.1$. The black curve is when the ratio is 100 and the curve peaks at $\pm5.3$. The curve in red has a ratio of 1000 and peaks at $\pm15.6$. The curves show that theoretically, if the ratio of load to ESR is large, the gain of the system improves.

From equation (3.117) the input inductor's parasitic resistance affects the gain more than the output inductor's parasitic resistance. The effects are shown in Figure 3-33 where the output inductor parasitic resistance was kept constant and the input inductor's parasitic resistance varied in factors of 10 from $700\mu\Omega$ to $70m\Omega$ ohms; the load resistor was kept constant at $100\Omega$ and output inductor's ESR was kept constant at $100m\Omega$.

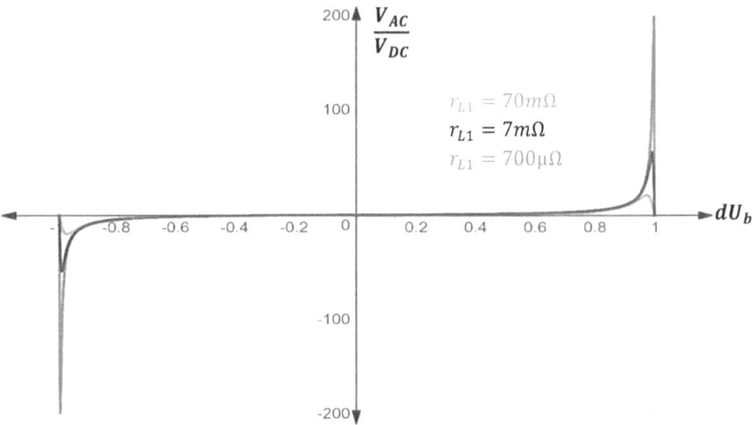


**Figure 3-33 Voltage transfer ratio at varying input inductor ESR and constant load and output inductor ESR**

Figure 3-33 shows that the gain drastically changes as the input inductor's ESR decreases. This is because the input current is larger than output current. This will lead to larger losses at the input ESR than at the output ESR. The output ESR was now varied from $1m\Omega$ in factors of 10 to $100m\Omega$; the load resistor kept constant at $100\Omega$ and input inductor's ESR constant at $70m\Omega$.

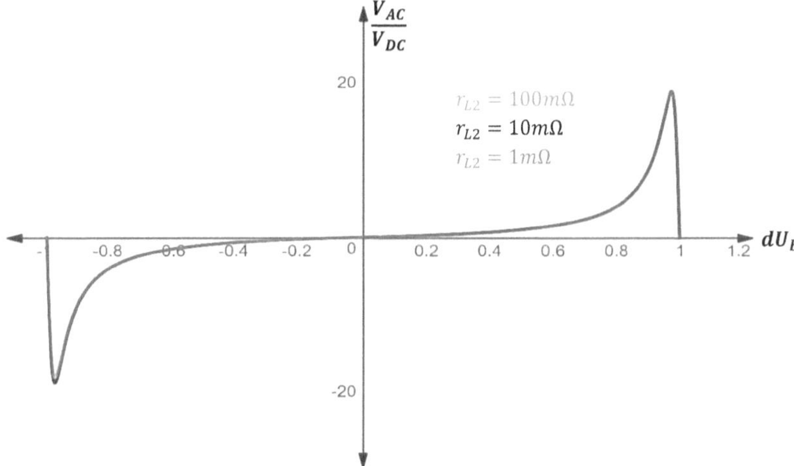


**Figure 3-34 Voltage transfer ratio at varying output inductor ESR and constant Load and input inductor ESR**

Figure 3-34 shows that all 3 curves are similar with almost no change as the output inductor ESR is varied. This was established from the equation derived.

## 3.7 Loss and efficiency analysis

The losses associated with the inverter can be attributed to the parasitic elements of the circuit. There are both switching and conduction losses.

### 3.7.1 Switch losses

The inverter has six switches that work together and the losses in each of them is discussed. The switches dissipate power during 4 stages: at switch ON, at switch OFF, during OFF-ON and during ON-OFF.

### i. Switching losses

During switch OFF-ON, there is an overlap as the voltage falls off and current rising and power is dissipated. Finally, during ON-OFF, as the voltage rises and the current decaying there is also an overlap which results in losses. During OFF-ON, the power losses were analysed [22].

$$P_{c(on)} = \frac{1}{2} V_{in} i_o f t_{c(on)} \tag{3.120}$$

For the converter stage switches $S_a$ and $S_b$ the frequency $f$ is $f_s$, during DC-AC power flow, $V_{in}$ is $V_{DC}$ and during AC-DC power flow, $V_{in}$ is $V_{AC,peak}$. For the unfolding bridge switches $S_c S_d S_e S_f$ the frequency is used $f_h$ and $V_{in}$ is $V_{AC,peak}$. Where $t_{c(on)}$ is comprised of the overlap between the voltage fall time and the current rise time. During the transition between ON-OFF state the losses are also given by the area subtended by the waveforms:

$$P_{c(off)} = \frac{1}{2} V_{in} i_o f t_{c(off)} \tag{3.121}$$

Where $t_{c(off)}$ is comprised of the overlap between the voltage rise time and the current rise time.

*ii.*     ***Conduction losses***

During the ON-state of switch, there is a small voltage which also constitutes losses:

$$P_{on} = V_{on} i_o \frac{t_{on}}{T} + i_o^2 r_{ds} \left( \frac{V_{in}}{V_o} \right)$$

(3.122)

Where $V_{on}$ is the switch's ON-state voltage; $r_{ds}$ the drain to source resistance. During switch OFF, if there is a small leakage current due to parasitic elements:

$$P_{off} = V_{in} i_{off} \frac{t_{off}}{T}$$

(3.123)

Where $i_{off}$ is the leakage current of the switch when it is OFF.

### 3.7.2    *Inductor losses*

For both the input and output inductors, the losses are given by:

$$P_L = i_L^2 r_L$$

(3.124)

Where $r_L$ is the individual inductor ESR resistance.

### 3.7.3    *Capacitor losses*

Both the input and output capacitors have:

$$P_L = i_L^2 r_L$$

(3.125)

Where $i_{c,RMS}$ is the RMS current across each of the input and output capacitors $C_1$ and $C_2$ respectively; $r_C$ the capacitors' ESR resistance.

### 3.7.4    *Efficiency analysis*

The expected efficiency during both AC-DC and DC-AC power flows was analysed and calculated using the equations derived.

#### i.    *DC-AC power flow inverter efficiency*

From Figure 3-4, the efficiency of the inverter can be determined by taking the output power divided by the input power or by subtracting the power loss from the input power:

$$e_{ff} = \frac{V_{dc}i_{dc} - P_{losses}}{V_{dc}i_{dc}} \tag{3.126}$$

The power losses were computed equations (3.120) – (3.125). Based on the specifications of the inverter and the switching parameters, the expected efficiency during DC-AC power flow using the equation above was calculated to be:

$$e_{ff} = 96.16\%$$

The efficiency value shown above is within the accepted standard for high efficiency switch mode power supplies.

#### ii.    *AC-DC power flow inverter efficiency*

During AC-DC power flow, the efficiency is the product of the diode bridge efficiency and the converter stage efficiency from Figure 3-13. The diode bridge efficiency was analysed in (3.86). From the coupling capacitor $C_1$ current:

$$\overline{i_{L2}} = -d\frac{\overline{i_{L1}}}{(1 - d)} \tag{3.127}$$

Where, in this case $\overline{i_{L2}}$ is equivalent to the continuous time-average rectified current drawn from the AC supply into the Ćuk converter. The average inductor current $\overline{i_{L1}}$ is the same as the DC load current required to charge the battery. Since during this power flow, the AC voltage is stepped down, the AC current is smaller than the DC current. Therefore, assuming a maximum battery charging current of $12A$ for a

$60V$ battery from a $311.23V$ peak AC source; the continuous-time averaged duty cycle $d$ is 0.16. Hence, the load current drawn from the supply is $1.62A$ RMS. The load $R_L$ seen by the rectifier is then $135.80\Omega$. The diode resistance $r_F$ is typically between $0.3 - 0.5\Omega$ and the worst case was chosen as $0.5\Omega$. The diode bridge efficiency is then:

$$Bridge_{eff} = \frac{2\sqrt{2}}{\pi}\left(\frac{135.80}{135.80 + 0.5}\right)\%$$

Which is an efficiency of 80.90%. The converter stage efficiency is calculated from the output power of the diode bridge – not the AC source – and subtracting the converter stage power losses:

$$Conv_{eff} = \frac{0.812V_{AC,rms}i_{AC,rms}cos\varphi\left(\frac{R_L}{R_L + r_F}\right) - \left(\overline{i_{L2}}^2 r_{L2} + (\overline{i_{L2}}^2 + \overline{i_{L1}}^2)r_{Fa} + \overline{i_{L1}}^2 r_{L1} + S_{b,switch,loss}\right)}{0.812V_{AC,rms}i_{AC,rms}cos\varphi\left(\frac{R_L}{R_L + r_F}\right)}$$

Where $cos\varphi$ is the input power factor (PF) which is assumed to be $-1$ for ideal reverse power flow; $r_{Fa}$ the DC resistance of the body diode of switch $S_a$. Substituting into expression, it evaluates to an efficiency of 97.95%. This efficiency is higher than the efficiency during DC-AC power flow since it does not include the bridge switches power losses. Therefore, the total efficiency is product of the bridge efficiency and converter stage efficiency:

$$e_{ff} = (97.95\% \times 80.90\%)$$

Which gives an efficiency of 79.24%. This efficiency is significantly lower than the expected efficiency during DC-AC power flow. This is because the diode losses are significantly higher than the switching losses.


## 3.8    Bidirectional Ćuk inverter dynamic analysis


To fully analyse the Ćuk Inverter's dynamics, the transfer function from output voltage to the input voltage was first be determined. This is because the output voltage must be controlled to control the power flow from DC to AC.

### 3.8.1 The bidirectional Ćuk inverter state space average model

The Ćuk inverter is a non-linear system due to the high frequency switching cycle. Controlling non-linear systems is challenging, therefore an averaged linear model is determined using state space averaging. The average model was then determined by using state space averaging. The state space was set up as follows:

When the input switch $S_a$ is ON:

$$X^{'} = A_1 X + B_1 V_{DC} \tag{3.128}$$

And when the switch is OFF:

$$X^{'} = A_2 X + B_2 V_{DC} \tag{3.129}$$

$$V_o = C_1 X \tag{3.130}$$

$$V_o = C_2 X \tag{3.131}$$

Using the state space equations above, the averaging models are described as:

$$X^{'} = [A_1 d + A_1(1 - d)]X + B_1 d + [B_2(1 - d)]V_{DC}$$

$$V_{AC} = [C_1 d + C_1(1 - d)]X$$

Where $d$ is the time-averaged continuous-time duty cycle. The dynamic equations that describe a Ćuk converter (inductor parasitic resistances and assuming the capacitive parasitic resistances are small enough to be neglected) when the input switch $S_a$ is closed are:

$$\frac{di_{L1}}{dt} = \frac{V_{DC}}{L_1} + \frac{r_{L1}i_{L1}}{L_1}$$

$$\tag{3.132}$$

$$\frac{di_{L2}}{dt} = \frac{V_{c1}}{L_2} = \frac{i_{L2} \parallel (r_{L2})}{L_2} - \frac{V_{c2}}{C_2}\left(\frac{1}{R} - 1\right)$$

$$\tag{3.133}$$

$$\frac{dVc_1}{dt} = -\frac{i_{L2}}{C_1}$$

$$\tag{3.134}$$

$$\frac{dVc_2}{dt} = -\frac{i_{L2}}{C_2} - \frac{V_{c2}}{RC_2}$$

$$\tag{3.135}$$

When input switch $S_a$ is open:

$$\frac{di_{L1}}{dt} = \frac{V_{DC}}{L_1} - \frac{V_{c1}}{L_1} + \frac{(r_{L1} + R)i_{L1}}{L_1}$$

$$\tag{3.136}$$

$$\frac{di_{L2}}{dt} = \frac{i_{c2}(r_{L1} + R)}{L_2} + \frac{V_{c2}}{L_2}\left(\frac{1}{R} - 1\right)$$

$$\tag{3.137}$$

$$\frac{dVc_1}{dt} = \frac{i_{L1}}{C_1}$$

$$\tag{3.138}$$

$$\frac{dVc_2}{dt} = \frac{Ri_{L2}}{RC_2} - \frac{V_{c2}}{RC_2}$$

$$\tag{3.139}$$

Where:

- $i_{L1}$ is the input current
- $i_{L2}$ is the output inductor current
- $V_{DC}$ is the input voltage
- $V_{c1}$ is the input capacitor voltage

- $V_{c2}$ *is the output capacitor voltage*
- $R$ is the DC or AC load resistor.

Using the equations (3.132) to (3.139) above and substituting them into the state space model equations (3.128) to (3.131), the matrices were set up as follows.

$$A_1 = \begin{bmatrix} -\dfrac{r_{L1}}{C_1} & 0 & 0 & 0 \\[2ex] 0 & -\dfrac{(r_{L2} \parallel R)}{L_2} & \dfrac{1}{L_2} & -\dfrac{1}{C_2}\left(\dfrac{1}{R}-1\right) \\[2ex] \dfrac{1}{C_1} & 0 & 0 & 0 \\[2ex] 0 & \dfrac{1}{C_2} & 0 & -\dfrac{1}{RC_2} \end{bmatrix} \tag{3.140}$$

$$A_2 = \begin{bmatrix} -\dfrac{r_{L1}}{L_1} & 0 & \dfrac{1}{L_2} & 0 \\[2ex] 0 & -\dfrac{(r_{L2} \parallel R)}{L_2} & 0 & -\dfrac{1}{C_2}\left(\dfrac{1}{R}-1\right) \\[2ex] \dfrac{1}{C_1} & 0 & 0 & 0 \\[2ex] 0 & \dfrac{1}{C_2} & 0 & -\dfrac{1}{RC_2} \end{bmatrix} \tag{3.141}$$

$$B_1 = B_2 = B = \begin{bmatrix} \dfrac{1}{L_1} \\[1ex] 0 \\[1ex] 0 \\[1ex] 0 \end{bmatrix} \tag{3.142}$$

$$C_1 = C_2 = C = \begin{bmatrix} 0 & 0 & 0 & 1 \end{bmatrix} \tag{3.143}$$

The open loop plant can be derived from the state space above using the equations below [41]:

$$P_o = C(SI - A)^{-1}B \tag{3.144}$$

$$X = V_{DC} C A^{-1} \tag{3.145}$$

Solving the equation above by substituting the state space constants gives:

$$P_o = \frac{d(r_{L1})s^3 + (dL_1 + C_2)s^2 + (r_{L1} + R)s + (L_1 r_{L1} + r_{L2} + C_1 R)}{(dL_1 L_2 + RC_2)s^4 + R(r_{L2}dL_1 + r_{L1}L_2)s^3 + (C_1 r_{L1} - dr_{L2}C_2)s^2 + (Rr_{L1}dr_{L2})s + (L_1 r_{L1} + r_{L2})} \tag{3.146}$$

The average model in (3.146) is shown to be linear. Controlling high order systems presents dynamic complexities to the control design. As such, several order reduction methods have been used to control the Ćuk converter. The other solutions were using current control methods instead. The issue with current control is that it has a complex control loop design. Furthermore, for the Ćuk converter, current control presents some instability issues due to the complexity of the converter. To control the Ćuk inverter, a new solution was presented in this study.

### 3.8.2 *Nested loop control strategy for the bidirectional Ćuk inverter*

The solution presented in this study employs a simpler, more effective alternative to the complex current control method which has complexities such as sub-harmonic oscillations and ramp compensation. Figure 3-35 shows that the control strategy splits the feedback control into 2 control loops: an outer voltage control loop and an inner current control loop based on the dynamic equations of the power transfer of the Ćuk converter, as analysed in 3.2.2ii.

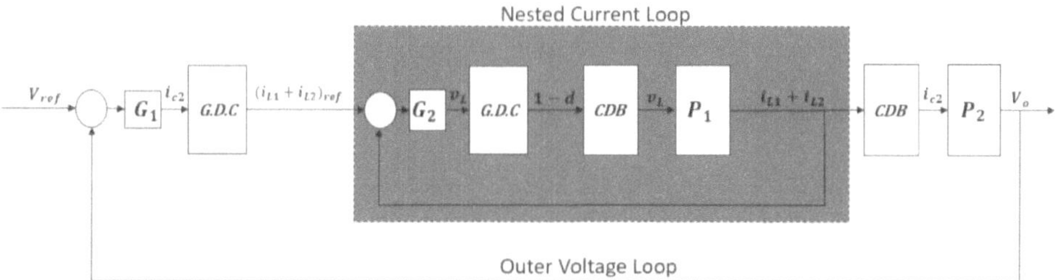


**Figure 3-35 Simplified nested loop control strategy model for the bidirectional inverter**

The goal of the control strategy is to split the model representing the inverter plant into two-linear plants $P_1$ and $P_2$, that have 1st order responses which can easily be controlled using classical control theory compensators $G_1$ and $G_2$, while being able to handle variable condition operation and reject complex input voltage and load currents disturbances. The Ćuk inverter transfers its power through the coupling capacitor, as such, the dynamic behaviour of the capacitor was used to analyse the dynamic behaviour of the inverter. The closed loop of the inverter is split as such because the duty cycle of the inverter presents non-linearities to the system. These non-linearities arise since both current and voltage depend on the duty cycle under variable conditions. The CDB blocks represent the Ćuk dynamic behaviour for either current or voltage and the G.D.C blocks represent the gain and disturbances compensations which will be discussed during the analysis of the individual loop analyses. The G.D.C block are compensations to cancel or minimise the effects of the CDB blocks which then reduces both inner and outer loops into simple classical control theory closed loop block diagrams but with linear low order plants. To control the inner current loop, the sum of two inductor currents was controlled; this is because the high frequency ripple across both inductors is the same, but the difference is in their average values. The average value of the inductor current $i_{L1}$ is the input current and the average of the inductor $i_{L2}$ is the output current. The duty cycle for the inverter is produced by the inductor current loop, as such, the control variable for the current loop was chosen to be the inductor voltage $v_L$ to eliminate the non-linearity arising from the loop gain, as will be discussed in detail during the design of the inner current control. The plant of this current loop reduces to a simple 2nd order system, based on the inductor's dynamic equation that can be compensated for using classical control theory methods like the root locus and bode plots. The output voltage is the same as the voltage appearing across the filter capacitor $C_2$, as such, the control input chosen for the output voltage was the current through the filter capacitor $C_2$ to eliminate the non-linearities arising from the duty cycle and this will also be discussed in detail during the voltage loop design. The plant of the voltage loop reduces to a 1st order system based on the output capacitor dynamic equation. Because of the symmetry of the Ćuk converter, the control strategy is applicable for both DC-AC and AC-DC power flows. During DC-AC power flow, $V_{in} = V_{DC}$ and $V_o = V_{AC}$. During AC-DC power flow, $V_{in} = V_{AC}$ and $V_o = V_{DC}$.

*i.*       ***Inner current loop dynamic modelling***

The goal of the design is to derive the dynamic behaviour of the Cúk inverter current loop and then subsequently, design the compensations for this behaviour, to control only a plant that is the inductor dynamic transfer function from its voltage to its current. Equation (3.147) shows the coupling capacitor $C_1$ voltage dynamic transient voltage transfer equation:

$$(1 - d(s))(v_{L2}(s) + V_0(s)) = d(s)(V_{in}(s) - v_{L1}(s)) \tag{3.147}$$

Expanding using the distributive law, equation (3.147) becomes:

$$v_{L2}(s) + V_0(s) - d(s)v_{L2}(s) - d(s)V_0(s) = d(s)V_{in}(s) - d(s)v_{L1}(s) \qquad \textbf{(3.148)}$$

From the theory development during section 3.2.2ii it was established that:

$$v_{L2}(s) = v_{L1}(s)$$

For any mode of operation for the bidirectional inverter. Using this fact, the equation (3.148) then simplifies to:

$$v_L(s) = d(s)\big(V_{in}(s) + V_0(s)\big) - V_0(s) \qquad \textbf{(3.149)}$$

Equation (3.149) shows the output voltage as independent of the duty cycle. This means that it appears as a disturbance in the loop. This is not ideal since the variations of the input voltage are not accounted for. Therefore, to compensate for input voltage variances, equation (3.147) was rearranged as shown:

$$V_0(s) = \frac{d(s)}{1 - d(s)}\big(V_{in}(s) - v_L(s)\big) - v_L(s) \qquad \textbf{(3.150)}$$

Substituting (3.150) into (3.149):

$$v_L(s) = d(s)\big(V_{in}(s) + V_0(s)\big) - \frac{d(s)}{1 - d(s)}\big(V_{in}(s) - v_L(s)\big) + v_L(s) \qquad \textbf{(3.151)}$$

Rearranging (3.151) by multiplying thorough by $\frac{1-d(s)}{d(s)}$:

$$v_L(s) = V_{in}(s) - \big(1 - d(s)\big)\big(V_{in}(s) + V_0(s)\big) \qquad \textbf{(3.152)}$$

Equation (3.152) now shows the input voltage as an input disturbance in the loop. This equation describes the Ćuk dynamic behaviour (CDB) for the inner current loop as shown in Figure 3-36. The inductor dynamic equation is given by (3.153):

$$v_L(s) = r_L i_L(s) + L s i_L(s) \tag{3.153}$$

From (3.153), solving for the inductor current:

$$i_L(s) = \frac{v_L(s)}{r_L + sL} \tag{3.154}$$

As stated in the design introduction, the goal for the inner current loop is to ensure that the sum of both inductor currents is controlled. This will ensure a more robust closed loop. As such, for both inductors (3.154) becomes:

$$i_{L1}(s) + i_{L2}(s) = \left( \frac{v_{L1}(s)}{r_{L1} + sL_1} + \frac{v_{L2}(s)}{r_{L2} + sL_2} \right) \tag{3.155}$$

Since both the inductor voltages are the same as already analysed, (3.155) becomes.

$$i_{L1}(s) + i_{L2}(s) = v_L(s) \left( \frac{1}{r_{L1} + sL_1} + \frac{1}{r_{L2} + sL_2} \right) \tag{3.156}$$

Substituting (3.152) into (3.156):

$$\begin{aligned}
i_{L1}(s) + i_{L2}(s) = {} & V_{in}(s) \\
& - (1 - d(s))(V_{in}(s) + V_0(s)) \left( \frac{1}{r_{L1} + sL_1} + \frac{1}{r_{L2} + sL_2} \right)
\end{aligned} \tag{3.157}$$

The sum of the inductor currents is seen as the output to the inner current loop. Since the inductor current depends on the output voltage – which is a variable gain, (3.157) becomes non-linear because of the duty cycle. To ensure that this non-linear characteristic is eliminated, the inductor voltage was chosen as the control input to the plant instead of the variable loop gain $(V_0(s) + V_{in}(s))$. This ensures that the plant is linear and only depends on the inductor's dynamic equation. The input voltage $V_{in}$ appears as an input disturbance. This can be compensated for by subtracting the disturbance in the loop gain but since the inner loop has a high bandwidth, this variable gain can be compensated by adding an inverse of the sum – these two together are the G.D.C blocks for the inner current loop. From (3.152) the duty cycle can be obtained as:

$$1 - d(s) = \frac{V_{in}(s) - v_{Lref}(s)}{V_o(s) + V_{in}(s)} \tag{3.158}$$

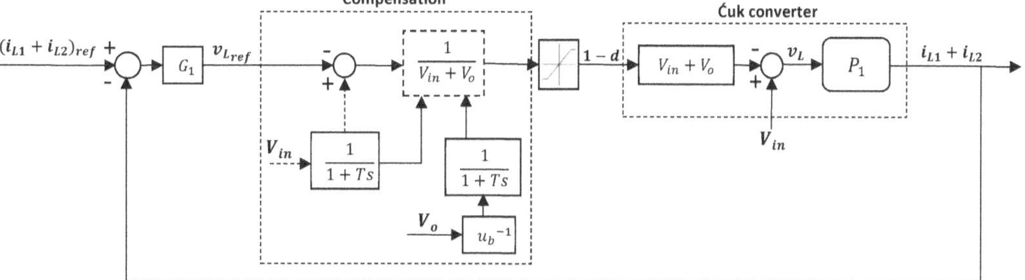


**Figure 3-36 Current control closed loop model**

The voltage loop shows how the duty cycle is generated and how feedforward works to compensate for the input voltage disturbance within the bandwidth of the current loop. The plant being controlled is now the linear transfer function from the inductor voltage to the inductor currents for both the inductors. Then the plant transfer function $P_1$ from (3.156) is:

$$P_1 = \frac{i_{L1}(s) + i_{L2}(s)}{v_L(s)} = \left(\frac{1}{r_{L1} + sL_1} + \frac{1}{r_{L2} + sL_2}\right) \tag{3.159}$$

Simplifying (3.159) then gives:

$$P_1 = \frac{i_{L1}(s) + i_{L2}(s)}{v_L(s)} = \left(\frac{s(L_1 + L_2) + (r_{L1} + r_{L2})}{(L_1 L_2)s^2 + (r_{L2} L_1 + r_{L1} L_2)s + (r_{L1} r_{L2})}\right) \tag{3.160}$$

Which is the plant transfer function for the inner loop with a high bandwidth. The Saturation block limits the duty cycle to between 0 minimum and 1 maximum. The $U_b$ blocks are as analysed in 3.2.2iii and defined by (3.17); they are required for AC power conversion systems. Purely DC-DC Ćuk power conversion systems don't require this block.

*ii.*      ***Outer voltage loop dynamic model***

Since the output voltage must be controlled – and is the same as the capacitor $C_2$ voltage – the coupling capacitor $C_1$ dynamic current equation was analysed:

$$1 - d(s)i_{L1}(s) = d(s)i_o(s) \tag{3.161}$$

But since the output current $i_o$ contains a high frequency ripple component filtered out by the output capacitor $C_2$, the equation becomes:

$$(1 - d(s))(i_{L1}(s)) = d(s)(i_{L2}(s) - i_{c2}(s)) \tag{3.162}$$

The dynamic equation describing the capacitor was analysed as follows:

$$i_{c2}(s) + r_{c2}C_2 i_{c2}(s) = C_1 s V_o(s) \tag{3.163}$$

Solving (3.162) for the output filter capacitor current $i_{c2}(s)$ and then substituting (3.162) into (3.163) and solving for the output voltage:

$$V_o(s) = \frac{i_{L2}(s) - \left(\dfrac{1 - d(s)}{d(s)}\right)i_{L1}(s) + r_{c2}C_2 i_{L2}(s) - \left(\dfrac{1 - d(s)}{d(s)}\right)i_{L1}(s)}{C_1 s} \tag{3.164}$$

The equation (3.164) shows the variable gain $\left(\frac{1 - d(s)}{d(s)}\right)$ which makes the open loop non-linear. To ensure the non-linearity is eliminated, the capacitor current $i_{c2}$ is used as a control input to the voltage loop plant to obtain the output voltage. Therefore, from (3.162):

$$i_{c2}(s) = i_{L2}(s) - \left(\frac{1 - d(s)}{d(s)}\right)i_{L1}(s) \tag{3.165}$$

From equation (3.165), $i_{c2}(s)$ is the input to the output voltage loop plant; but the reference of the sum of the two inductor currents can't be obtained with the equation in that form. Therefore, on the right-hand side, an inductor current subtraction, redundant term was added to rephrase the equation:

$$i_{c2}(s) = i_{L2}(s) - \left(\frac{1 - d(s)}{d(s)}\right)i_{L1}(s) + (i_{L1}(s) - i_{L1}(s)) \qquad (3.166)$$

From (3.161), solving for the input inductor current $i_{L1}(s)$:

$$i_{L1}(s) = \left(\frac{d(s)}{1 - d(s)}\right)i_o(s) \qquad (3.167)$$

Therefore substituting (3.167) into (3.166):

$$i_{c2}(s) = \left(i_{L2}(s) + i_{L1}(s)\right) - i_o(s) - \left(\frac{d(s)}{1 - d(s)}\right)i_o(s) \qquad (3.168)$$

Simplifying (3.168):

$$i_{c2}(s) = \left(i_{L2}(s) + i_{L1}(s)\right) - i_o(s)\left(\frac{1}{1 - d(s)}\right) \qquad (3.169)$$

Factoring the duty cycle term out of (3.169) and rearranging; the input to the output voltage loop plant, $i_{c2}(s)$ is:

$$i_{c2}(s) = \frac{1}{1 - d(s)}\left(\left(i_{L2}(s) + i_{L1}(s)\right)(1 - d(s)) - i_o(s)\right) \qquad (3.170)$$

Equation (3.170) describes the voltage loop Ćuk dynamic behaviour (CDB). This is illustrated in the outer loop block diagram in Figure 3-37:

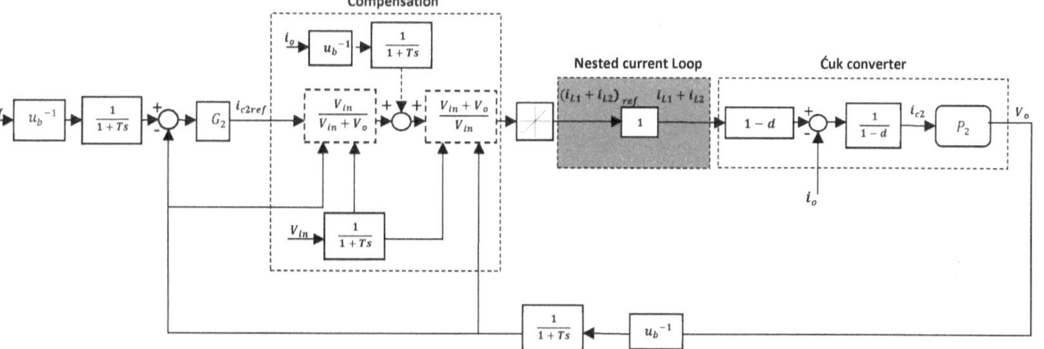


Figure 3-37 Voltage control closed loop model

As a consequence of representing the equation as in (3.170), the voltage control loop CDB in Figure 3-37 shows 2 loop gains: $1 - d$ and its inverse. These gains make the open loop system non-linear and furthermore, the voltage loop has much slower dynamics than these loop gains since i.e. the duty cycle switches states at $20kHz$ and the voltage loop only at $50Hz$ in DC-AC control or $0Hz$ in the case of DC-DC control. As such, they cannot be compensated for completely – but since the Ćuk inverter has small inductor sizes, the inductor variations can be neglected and equation (3.147) can be approximated by:

$$\frac{d(s)}{1 - d(s)} \approx \frac{V_o(s)}{V_{in}(s)} \tag{3.171}$$

From Figure 3-35, the inductor current sum is required as a reference for the current loop. Substituting (3.171) into (3.168) gives and rearranging:

$$\left(i_{L2}(s) + i_{L1}(s)\right)_{ref} = i_{c2,ref}(s) + i_o(s) + \left(\frac{V_o(s)}{V_{in}(s)}\right) i_o(s) \tag{3.172}$$

Simplifying (3.172):

$$\left(i_{L2}(s) + i_{L1}(s)\right)_{ref} = i_{c2,ref}(s) + i_o(s) \left(\frac{V_{in}(s) + V_o(s)}{V_{in}(s)}\right) \tag{3.173}$$

Rearranging the equation:

$$\left(i_{L2}(s) + i_{L1}(s)\right)_{ref} = \frac{V_{in}(s) + V_o(s)}{V_{in}(s)} \left(i_{c2,ref}(s)\frac{V_{in}(s)}{V_{in}(s) + V_o(s)} + i_o(s)\right) \tag{3.174}$$

The equation (3.174) shows two slow dynamic loop gains, the first:

$$Loop\ Gain1 = \frac{V_{in}(s) + V_o(s)}{V_{in}(s)} \tag{3.175}$$

The loop gain in (3.175) compensates for the $1 - d$ loop gain. This means that the inverse $1 - d$ loop gain can be compensated for by (3.176):

$$Loop\ Gain\ 2 = \frac{V_{in}(s)}{V_{in}(s) + V_o(s)} \tag{3.176}$$

These loop gains don't fully cancel out the effects of the duty cycle gains but provide the best compensations for them as shown in Figure 3-37. They can compensate for the duty cycle loop gains within the bandwidth of the outer voltage loop. The output current's disturbance can be accounted for by removing it at the output as shown by the loop. This compensation acts as feed-forward control to ensure that any sudden changes in the output current can be compensated for. This compensation plus the two loop gains *Loop Gain* 1 and *Loop Gain* 2 are the G.D.C block for the voltage loop. The saturation block shown limits the current sum of the input and output inductors from $100A$ maximum to $-50A$ minimum for the nominal operation; these limits can be varied depending on the test being done on the inverter; this is to protect the switches against over-currents and inrush currents. The diagram shows the plant being controlled is now given by the linear transfer function from the output capacitor current to the output capacitor voltage. From (3.163):

$$P_2 = \frac{V_o(s)}{i_{c2}(s)} = \frac{sC_2 + (1 + r_{c2})}{sC_2} \tag{3.177}$$

Which is the plant transfer function of the outer voltage loop

# 4.  Bi-directional Ćuk inverter design

The Bi-directional inverter was designed for using the analysis done in the previous section. The specifications for the design were relayed. Then the component sizing of all the passive elements together with the selection of the active components. The dead time to be introduced in the system was designed for and the practical processor and transducers were selected. Finally, the control loop compensators were designed for.

## 4.1    Design Specifications

The specifications required for the inverter were based on the EN 501060 standard as reviewed in section 2.1.3. Figure 4-1 shows the overall system being designed for.

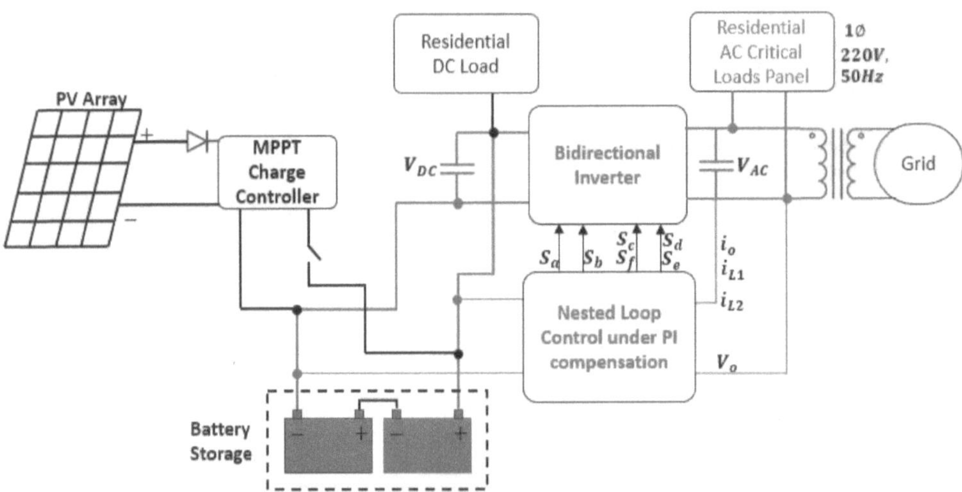


**Figure 4-1 Design block diagram**

The DC power supply mimics the battery as the power source. It was also used a DC load which would mimic the requirements for charging a battery. The inverter specifications were detailed in Table 4-1.

Table 4-1 Bi-directional inverter specifications

| | |
|---|---|
| Battery power output | $1\ kW$ |
| Battery / DC output voltage | $60 - 70V_{DC}$ |
| Grid / AC output voltage | $220V$ |
| Grid fundamental frequency, $f_h$ | $50H_z$ |
| Carrier switching frequency, $f_s$ | $20kH_z$ |
| DC load impedance, $R_{Lb}$ | $5\Omega \leq R_{Lb} \leq 20\Omega$ |
| AC load impedance, $R_L$ | $50\Omega \leq R_L \leq 300\Omega$ |
| Inductor ESR, $r_L$ | $50m\Omega$ |
| Capacitor ESR, $r_C$ | $30m\Omega$ |
| Output voltage ripple, $\Delta V_o$ | $\leq 2\%\ of\ V_o$ |
| Coupling capacitor voltage ripple, $\Delta V_{C1}$ | $5\%\ of\ V_{DC}$ |
| Inductor current ripple, $\Delta i_L$ | $25\%\ of\ i_{L,max}$ |
| Output voltage load regulation | $\pm 10\%\ of\ V_o$ |
| Output voltage line regulation | $\pm 5\%\ of\ V_o$ |
| %THD | $< 8\%$ |
| Maximum duty cycle, $d_{max}$ | $83.69\%$ |
| Minimum duty cycle, $d_{min}$ | $16.30\%$ |

The DC Load impedance was chosen such that a $60 - 70V_{DC}$ battery has a maximum charge current of 12A of which, the equivalent resistive load is 5Ω and the minimum charge current is 3A which is load of 20Ω. The AC load was designed for load taking between $1 - 6A$ peak current – which is $0.91 - 4.54A\ RMS$ rated current. This would be the rated current for the critical loads panel for this 1kW system. This system would power 6, 100W parallel connected incandescent light bulbs, with each consuming 0.45A of current; a 138W plasma TV consuming 0.63A of current and a ceiling fan set to max speed consuming 0.6A of current.

## 4.2    Battery selection

Based on the literature done, the battery which would suit the design is the Lithium ion battery. This is because they have the highest terminal voltage and energy density of the batteries reviewed. The Battery Chosen was the Model: LIR18650 2600mAh $60V, 20Ah$ whose datasheet can be found in the appendix.

The open circuit voltage for the battery is 70V. The battery has an internal resistance of $320m\Omega$ therefore, with a maximum charging current of $12A$, the charging voltage required by the battery is the open circuit voltage plus the voltage drop across the internal resistance i.e. $70 + (12 \times 0.32)$, which is a charging voltage of $73.84V$. This was used as the voltage reference during AC-DC power flow testing.

## 4.3     Passive components design

Both the input and output inductors, the coupling, decoupling and filter capacitors, from Figure 3-1, were designed for and detailed in this section.

### 4.3.1     Input Inductor design

The inverter is to be designed such that during DC-AC power flow the inductor operates in CICM and during AC-DC power flow, DCVM. As such, there have to exist a set of values within the minimum inductor value required to ensure CICM and the maximum inductor value required for DCVM – as analysed in sections 3.3.1i and 3.3.2i respectively. This set intersection is defined by:

$$L_{1min} \leq L_1 < L_{1max} \tag{4.1}$$

Substituting the equations (3.35) and (3.55) into (4.1):

$$\frac{R_{L,min}(1 - d_{min})^2}{2d_{min}f_s} \leq L_1 < \frac{1}{C_{2,min}f_s^2} \tag{4.2}$$

Substituting the values from Table 4-1, the number line in Figure 4-2 was drawn:

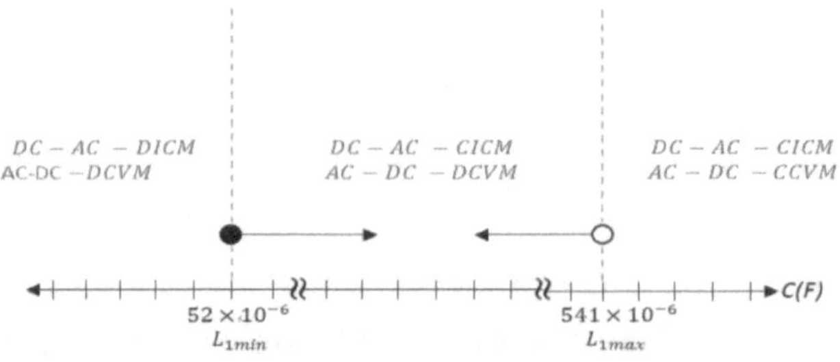


**Figure 4-2 Input inductor region of selection**

The number line shows that there exists an intersection such that the operation of the inverter can be realised. To compute this input inductor value, the ripple input inductor current should be considered. This was chosen to be 25% of the maximum inductor current. Based on Figure 3-8 (b), the average input inductor $\overline{i_{L1}}$ can be computed as:

$$\overline{i_{L1}} = i_{L1,max} - \frac{i_{L1,max} - i_{L1,min}}{2} \tag{4.3}$$

Since the numerator of the second term in (4.3) is the ripple input inductor current $\Delta i_{L1}$, therefore, the equation becomes:

$$\overline{i_{L1}} = i_{L1,max} - \frac{0.25 i_{L1,max}}{2} \tag{4.4}$$

Solving (4.4) for the maximum input inductor current:

$$i_{L1,max} = \frac{8\overline{i_{L1}}}{7} \tag{4.5}$$

The average input inductor current $\overline{i_{L1}}$ is the same as the maximum input discharge current $i_{DC}$, which, based on Table 4-1, is 16.67A; from this, the inductor size was computed from

$$\Delta i_{L1} = \frac{d_{max} T_s (V_{DC} - r_{L1} i_{L1,max})}{2L_1} \tag{4.6}$$

Rearranging gives:

$$L_1 = \frac{d_{max} T_s (V_{DC} - r_{L1} i_{L1,max})}{2\Delta i_{L1}} \tag{4.7}$$

Using the Table 4-1 and substituting into the equation the inductor was computed to be $501 \mu H$. This value is within the set of values required. The practical inductor used was built a ferrite core found in the Appendix. This wire used was a $5mm$ diameter copper wire to reduce the DC inductor resistance. The inductor was assemble using the equation:

$$L_1 = \frac{N^2 \mu_o \mu_r A_c}{l} \tag{4.8}$$

Where $N$ is the number of copper wire turns around the core; $\mu_o$ is the permeability of free space; $\mu_r$ is the relative permeability of the core; $A_c$ the cross-sectional area of the core and $l$ the length of the copper wire used. The length of the wire, $l$ is the unknown variable required to calculate the number of turns. To obtain the length, the resistor formula was used.

$$r_{L1} = \rho \frac{l}{A_l} \tag{4.9}$$

Substituting into the equations (4.9) into (4.8), the inductor used was measured to be $505\mu H$ at $20kH_z$ and the DC resistance measure was $70m\Omega$. The practical values were slightly higher than those desired from the specification but were still within the limitations of the design.

### 4.3.2 Output inductor design

The region of selection for the output inductor was also determined using the condition:

$$L_{2min} \leq L_2 < L_{2max} \tag{4.10}$$

Substituting equations (3.38) and (3.56) into (4.10):

$$\frac{R_{Lmin}(1 - d_{min})}{2f_s} \leq L_2 < \frac{1}{C_2 f_s^2} \tag{4.11}$$

Figure 4-3 shows the number line representing the intersection set of values from within the output inductor values must be selected.

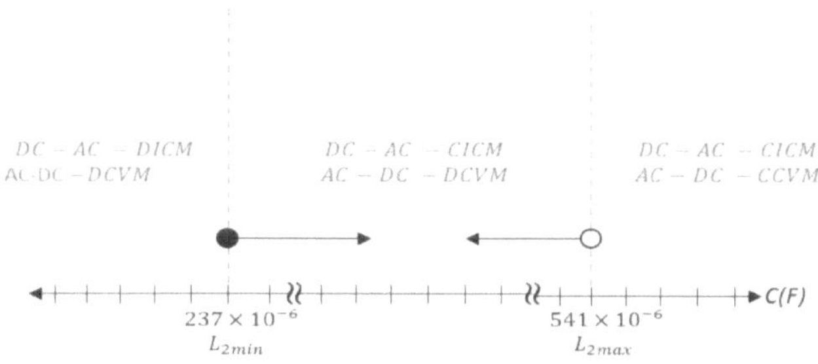


**Figure 4-3 Output inductor region of selection**

The number line shows that the minimum value for the output inductor is much higher than that of the input inductor value; this is because the AC current is much smaller than the current drawn from the battery. The output inductor was design in the same manner as the input inductor. The same 25% ripple factor was chosen but for the maximum output inductor $i_{L2,max}$ which using Table 4-1 and Figure 3-8 (b) is 4.63A and using the equation (4.12):

$$\Delta i_{L2} = \frac{(1 - d_{max})T_s\left(V_{AC,peak} - r_{L2}i_{L2,max}\right)}{2L_2} \qquad (4.12)$$

Rearranging (4.12) gives:

$$L_2 = \frac{(1 - d_{max})T_s\left(V_{AC,peak} - r_{L2}i_{L2,max}\right)}{2\Delta i_{L2}} \qquad (4.13)$$

Using the specifications and substituting into the equation the output inductor was found to be $504\mu H$. After the practical assembly of the inductor the value measured was, $510\mu H$ at $20kHz$ and the DC resistance was $110m\Omega$.


### 4.3.3 *Input coupling capacitor design*

The input capacitor is responsible for the energy transfer in the inverter. During CICM the inductor currents and coupling capacitor voltage are continuous; but during DCVM, the coupling capacitor voltage is discontinuous – although the inductor currents are still continuous. This selection is an engineering trade-off since to ensure discontinuity in $V_{C1}$, the coupling capacitor $C_1$ must be as small as possible, but this capacitor also must be large enough such that its ripple voltage $\Delta V_{C1}$ is negligible from its average value during CICM.

From the analysis in section 3.3.1iii, the minimum coupling capacitor equation was determined; in section 3.3.2, the maximum coupling capacitor to ensure DCVM was determined. Therefore, from the condition in equations (3.46) and (3.49):

$$C_{1min} \leq C_1 < C_{1max} \tag{4.14}$$

Substituting for the minimum and maximum values of the duty cycle:

$$\frac{d_{min}}{2f_sR_{L,max}} \leq C_1 < \frac{d_{max}^2}{2f_sR_{Lb,min}} \tag{4.15}$$

The condition in states that, for the bi-directional inverter to operate as desired, there must exist a range of coupling capacitor values such that the desired coupling capacitor value lies within that range i.e. the two inequalities must have an intersection. The values were substituted, and the inequality evaluated. The results were shown in .

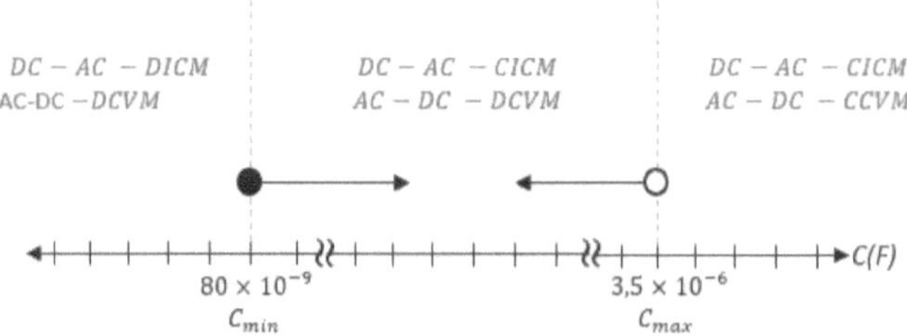


Figure 4-4 Coupling capacitor region of selection

By inspection from the number line and the min and max values, the intersection of values exists. The capacitor value can be calculated from:

$$C_1 = \frac{V_{AC,peak}d_{max}T_s}{R_{Lmax}\Delta V_{c1}} \tag{4.16}$$

Substituting value from Table 4-1, the coupling capacitor was found to be $0.64\mu F$ which large enough that the capacitor voltage is constant and small enough to ensure DCVM operation. The capacitor value used was 0.44 due to the availability of high voltage capacitors in the lab. This value was still within the required specification range. This selection is an engineering trade-off since to ensure discontinuity in

$V_{C1}$, the coupling capacitor $C_1$ must be as small as possible, but this capacitor also must be large enough such that its ripple voltage $\Delta V_{C1}$ is negligible from its average value during CICM.

### 4.3.4    Output filter capacitor design

The ripple at the output is an important aspect of the design. It is represented as a percentage of the average value of the output voltage. The illustration of voltage ripple is shown in Figure 4-5.

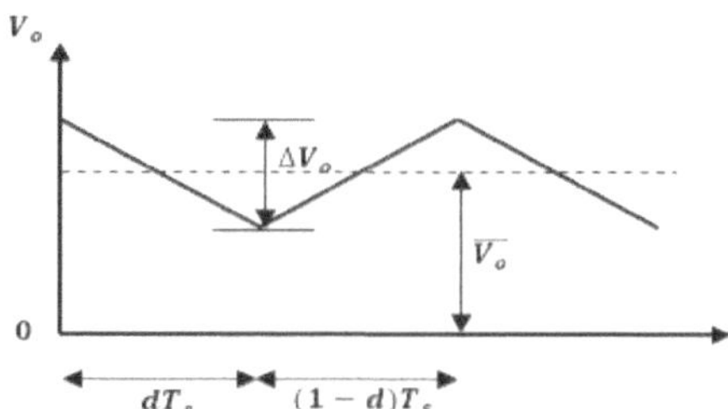


**Figure 4-5 Output voltage ripple of a Ćuk converter**

$\Delta V_o$ is the output voltage ripple. It is desirable for it to be as small as possible. If the output capacitor chosen is large enough – assuming a constant load resistor – the Figure 4-6 shows that the ripple will be low because the time constant will be high i.e. it will take a long time for the capacitor to charge up.

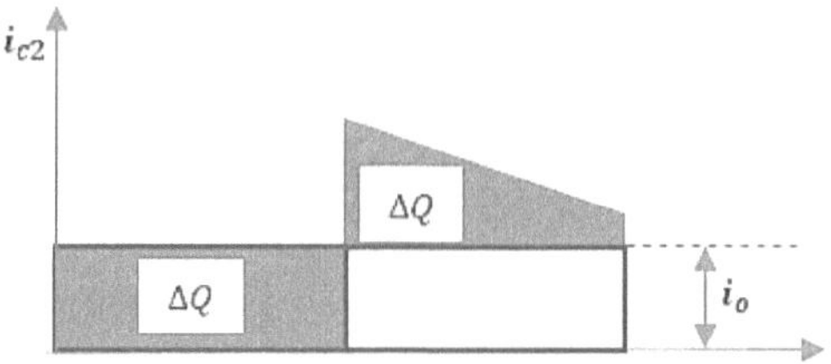


**Figure 4-6 Charge across the output capacitor over a single period**

The charge across the capacitor is represented by $\Delta Q$ and its distribution over a period. The charge during the OFF and ON states is equal. To evaluate, only 1 state needs to be considered.

$$\Delta V_{c2} = \frac{1}{C_2}\left(\int_{o}^{dT_s} i_{C2}\, dt\right) \qquad (4.17)$$

Evaluating the integral and using the fact that:

$$i_{AC} = \frac{V_{AC}}{R_L} \tag{4.18}$$

And the output is constant, the following is obtained:

$$\Delta V_c = \frac{dT_s V_{AC}}{R_{L,min} C_2} \tag{4.19}$$

This equation (3.19) can then be used to work out the appropriate capacitance to achieve the desired output voltage high frequency ripple. The ripple is 2% and substituting into the equation using the specification in Table 4-1, the output capacitor value was found to be $27\mu F$. An aluminium capacitor was chosen whose capacitance measures at $30\mu F$.

### 4.3.5  *Decoupling Capacitor*

The voltage at the supply fluctuated as a $100Hz$ current drawn from the supply. As such, a decoupling capacitor was added at the DC input to minimise the ripple of the current. The fluctuations in the input voltage is given by [42]:

$$C_d = \frac{V_{DC,rated} i_{DC,rated}}{2\pi f V_{DC,rated} \Delta V_{c1}} \tag{4.20}$$

Substituting for values from Table 4-1, the decoupling capacitor value was found to be $3.18mF$. The value used was $4mF$, which was two $2mF$ capacitors connected in parallel.

## 4.4    Active switch design selection

The switch selected needed to be able handle high voltages and gate drain currents while operating at a mid-range frequency of 20 kHz. The switch was selected based on how much rise and fall times it has because this influences the dead time and how much voltage stress it can handle.

### 4.4.1   Voltage stress

Based on the analysis done, the highest stress voltage would be observed under the DCVM. Substituting the values into equation (3.42) shows a stress voltage of at least $750V$. The switch selected was to have a voltage rating above this value.

### 4.4.2   Dead time

Based on the voltage stress above the switch chosen was the IXXH110N65C4 found in the appendix. It has a drain to source voltage of 900V and can handle 110A of current from the drain to the source. From Figure A14.2 in the Appendix A14, the following equations were derived.

$$T_{gsp} = 4C_{iss}\left(\frac{V_{gs} - V_{gp}}{I_{goff}}\right) \tag{4.21}$$

$$T_{gpt} = R_{goff}\left(\frac{Q_{gd}}{V_{gp}}\right) \tag{4.22}$$

$$T_{dsd} = \frac{\pi}{2}\sqrt{\frac{L_{PCB} \times Q_{OSS}}{V_{IN}}} \tag{4.23}$$

$$T_{lsh} = t_{rmax} - t_{fmax} \tag{4.24}$$

The dead time needs to be greater than all these times to ensure that the switch is either fully ON or fully OFF. Hence, the dead time is:

$$T_{dt} \geq T_{gsp} + T_{gpt} + T_{dsd} + T_{lsh} \tag{4.25}$$

If the dead time is above these times, the switch will operate safely. Based on the dead time selection equation (4.25) the IGBT has the following minimum dead time:

$$T_{dt} \geq 2.65\mu s$$

This is the dead initialised on the gate drive circuits and MATLAB Simulink for simulations.


## 4.5 Transducers

To convert the electrical signals from the Inverter into signals that could be analysed electronically, transducers were used. The transducers were required to have a high bandwidth and a high level of noise rejection.


### 4.5.1 Voltage transducers

There were 2 voltages transducers needed for the input voltage and the output voltage. The bandwidth of the outer voltage loop is significantly low so there was no need for an active filter at the output of the transducer. One of the transducers needed to have a range of at least $0$ $to$ $80V$ and the other a range of at least $0$ $to$ $315V$. The Voltage Transducer selected was the LEM LV25P.


### 4.5.2 Current transducers

The other 3 transducers were for the input and output inductor currents and the output currents. Since the inner loop dynamics are much faster than the loop, a filter was required to realise the control signal. The ranges needed for the transducers were $-20$ $to$ $30A$ for the inductor currents and $0$ $to$ $6A$ for the output current. The transducer chosen was the LA55P.


## 4.6 PWM generator

The PWM signals for the IGBT switches were obtained using an external PWM generator shown in Figure 4-7. This is because the sampling of the DPSACE was incapable of handling the high switching frequency of the circuit.

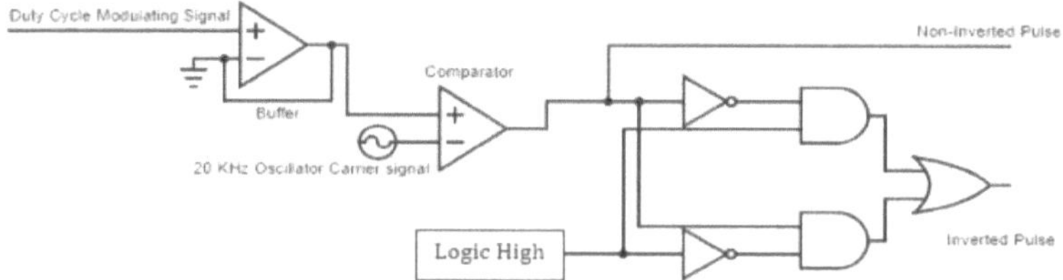


**Figure 4-7 Inverter PWM generator**

The generator takes in a modulating signal from the DSPACE's DAC pin 2 and compared it to the high frequency oscillator. The XOR gate then inverts one of the signals.

## 4.7    Nested loop control strategy compensator designs

The lead compensators for the nested current and voltage loops were designed in this section. The overall system being designed for is a 3rd order non-linear system but using the control loop design discussed in the analysis, there were plants being controlled of which the outer loop plant was 2nd order and the inner loop plant was 1st order and both were linear. The specifications were first detailed and then based on the specifications the compensators were designed.

### 4.7.1    Design specifications

The design specifications were tabulated as shown Table 4-2. The steady state error, bandwidth, settling time, phase and gain margins apply to both DC-AC and AC-DC power flows. The percentage overshoot was designed based on the charging voltage of the battery during AC-DC power flow.

**Table 4-2 Control Design Specifications**

| Specification Type | Specification |
|---:|---|
| Steady State Error | Zero |
| Percentage Overshoot | $\leq 17\%$ |
| Current loop Bandwidth | $\geq 2500Hz$ |
| Voltage Loop Bandwidth | $\geq 250Hz$ |
| Phase margin | $\geq 58^o$ |
| Gain Margin | $\geq 5dB$ |
| Settling time | Faster than $30ms$ |

The specifications in Table 4-2 were determined based on the following:

*i.*     ***Steady state error***

The loops include some gain values that vary, therefore any error in steady state would present unstable magnitudes for both DC-AC and AC-DC power flows.

*ii.*     ***Percentage overshoot***

The percentage overshoot was based on a single Li-ion cell. The nominal charging voltage a Li-ion cell is $3.6V$ and the maximum voltage it can withstand – for a short time – without damage, is $4.2V$. Based on the following equation:

$$\%OS \leq \frac{V_{max} - V_{nom}}{V_{nom}} \tag{4.26}$$

Where $V_{max}$ is the maximum charging voltage; $V_{nom}$ the nominal charging voltage; the overshoot in Table 4-2 was obtained.

*iii.*     ***Phase margin***

The phase margin ($PM$) was based on the percentage overshoot as shown by equation (4.27):

$$PM = 75 - \%OS \tag{4.27}$$

Substituting for the values gave the *PM* shown in Table 4-2.

### 4.7.2    *Current loop lead compensator design*

The transfer function being design for, based on the analysis in section 3.8.2i of the inverter:

$$P_1 = \frac{i_{L1} + i_{L2}}{v_L} = \left( \frac{s(L_1 + L_2) + (r_{L1} + r_{L2})}{(L_1 L_2)s^2 + (r_{L2}L_1 + r_{L1}L_2)s + (r_{L1}r_{L2})} \right)$$

The plant was simulated on MATLAB to show the bode, root locus and time responses.

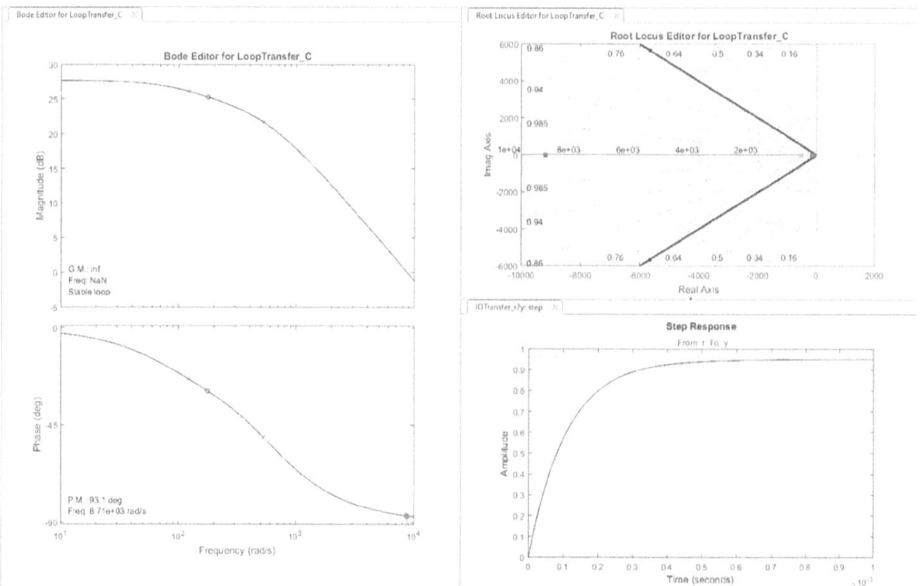

**Figure 4-8 Current plant closed loop step response**

They 3 plots show that the closed loop is stable. The root locus shows the region where the closed loop poles must lie to achieve the desired results. The closed loop poles start at open loop poles and terminate at open loop zeros. The closed loop poles are shown in pink. It also shows that the dominant pole and the only zero are placed next to each other on the real axis. The system then behaves based on the pole at $-61\ rad$ as the dominant pole.

The bode plot shows that even though the system is 2nd order, the response is like that of a 1st order system. This is because of the dominant pole as discussed above. The plot also shows that the gain margin is infinite, and the phase margin is $92^o$. The bandwidth achieved by the current loop is at $628Hz$ which does not meet the $2500Hz$ specification.

The time step response shows the system is critically damped. The settling time observed is $55ms$ which is not within the specification required. There is a small steady state error.

The current loop plant is a $TYPE$ 0 system, meaning that it will have finite steady state error if given a step input. During steady state, $T$ tends to infinity which makes "$s$" tend to zero. Therefore, using the final value theorem,

$$e_\infty(t) = e_o(s) = \lim_{s \to 0} s \frac{1}{1 + P_1(s)} R(s) \tag{4.28}$$

Since the input is a step response;

$$R(s) = \frac{1}{s}$$

Substituting for the $P_1$ into equation (4.28):

$$e_o(s) = \lim_{s \to 0} s \left( \frac{(L_1 L_2)s^2 + (r_{l2}L_1 + r_{l1}L_2)s + (r_{l1}r_{l2}) + s(L_1 + L_2) + (r_{l1} + r_{l2})}{(L_1 L_2)s^2 + (r_{l2}L_1 + r_{l1}L_2)s + (r_{l1}r_{l2}) + 1} \right) \frac{1}{s} \tag{4.29}$$

Evaluating the limit gives:

$$e_o(0) = \frac{(r_{l1} + r_{l2}) + (r_{l1}r_{l2})}{(r_{l1}r_{l2}) + 1} \tag{4.30}$$

Which is a finite steady state error. Substituting for ESR resistances gives a steady state error of 0.1 which is 10% of the required $220V\ RMS$ final value for the Inverter side and 10% of the $60V$ final value required for the rectification.

The lead circuit structure is shown in equation (4.31). It is needed to increase the bandwidth while keeping the system at zero steady state error.

$$Lead = \frac{a_2 \tau s + 1}{\tau s + 1} \tag{4.31}$$

From Figure 4-8, the lead compensator needs to add phase between $1\ rad/s$ and $2000\ rad/s$ which will also add some gain and increase the bandwidth. The amount of phase to be added is 40 degrees of phase at $100\ rad/s$. To determine the parameters of the lead compensator, the following equations were considered. The maximum phase:

$$\varphi_{max} = \sin^{-1}\left(\frac{a_2 - 1}{a_2 + 1}\right) \tag{4.32}$$

The maximum phase substituted was 40 plus an extra 15 degrees added as part of a lead circuit design safety factor. After obtaining $a_2$ using the fact that this phase will be added at $100\ rad/s$:

$$\omega_m = \frac{1}{\tau\sqrt{a_2}} \tag{4.33}$$

In equation (4.33) $\omega_m$ is $100\ rad/s$ - then $\tau$ was solved for by substituting into (4.31) results from equations (4.32) and (4.33), The lead compensator $G_1$, obtained was:

$$G_1 = \frac{0.0047s + 1}{0.00046s + 1}$$

The compensator designed was in fact, a lead compensator since $a_2 > 1$. After simulating the closed loop, the system showed a slight lag in speed and the proportional controller gain was designed to increase the speed. The root locus was used to design for this. The root locus characteristic equation can be described by:

$$P = \frac{N(s)}{D(s)} \tag{4.34}$$

The characteristic equation is:

$$D(s) + KN(s) = 0 \tag{4.35}$$

Therefore, substituting in $Plant_1$ into gives:

$$(L_1 L_2)s^2 + (r_{l2}L_1 + r_{l1}L_2)s + (r_{l1}r_{l2}) + Ks(L_1 + L_2) + K(r_{l1} + r_{l2}) = 0$$

The dominant pole position on the root locus to achieve the desired settling time is:

$$s = -123.145 \pm 98.71i$$

These closed loop pole positions lie at the edge of the desired region. Substituting for "$s$" and the electrical elements:

$$K = 5.8752$$

This gain was then added to the loop and following results were obtained. The root locus now shows the new position of the closed loop poles which have moved from open loop poles because of the proportional gain. These closed loop poles are at $s = -130.23 \pm 100.23i$ which is close to the desired pole locations.

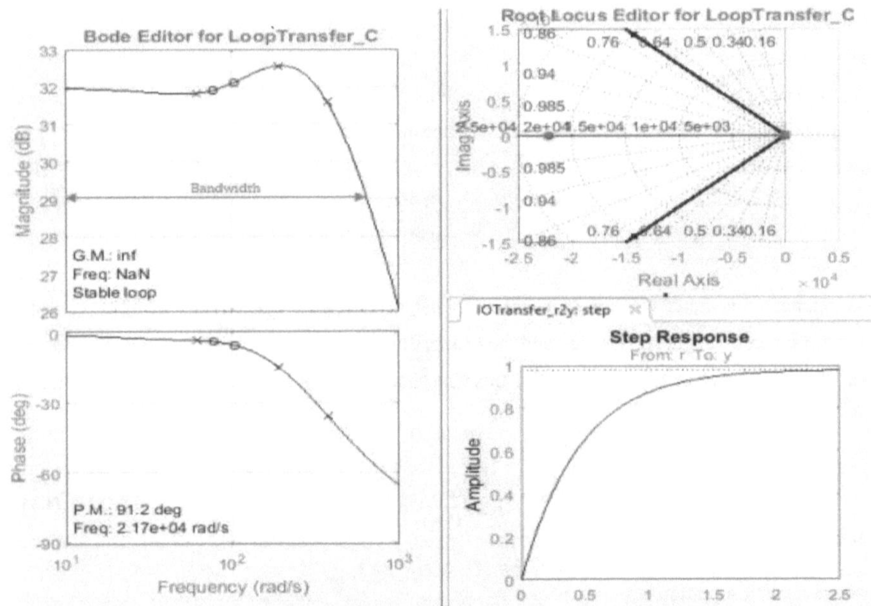

Figure 4-9 Lead compensation closed loop response for current loop

Figure 4-9 shows that the bandwidth of the system has now increased. The bandwidth observed above is $4140Hz$. The phase margin has also increased to 91 from 55. The overshoot of the system has now diminished. The speed has also increased to settling at $25ms$. The lead compensator has improved the system's dynamic response to meet all required specifications.

### 4.7.3    *Voltage loop lead compensator design*

The $Plant_2$ being controlled in this loop is as analysed in section 3.8.2ii:

$$P_2 = \frac{v_o}{i_{c2}} = \frac{1 + r_{c2} + sC_2}{sC_2}$$

The Plant is $TYPE$ 1 meaning that it will have zero steady state error when given a step input. The closed loop step response was also plotted for the outer loop.

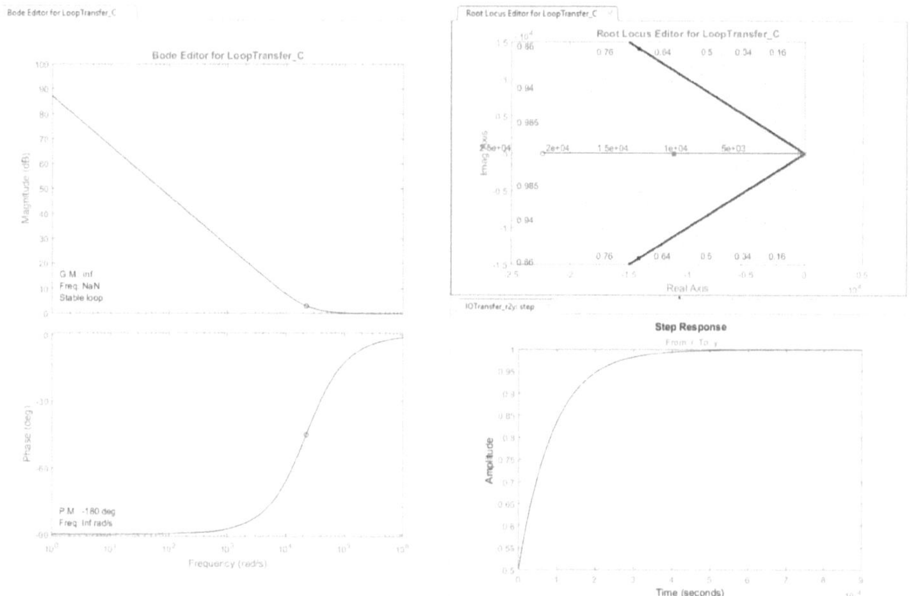

**Figure 4-10 Voltage transfer function closed loop step response**

Figure 4-10 shows that the system is stable. The root locus shows that the pole position is at the origin which no steady state error. The system shows that the settling time is $0.6ms$ which considerably fast.

The system is also critically damped with no overshoot. The gain margin on the Bode plot is shown to be infinite and the phase margin is $90^o$. The bandwidth of the voltage loop plant is $5Hz$ which does not meet the specifications. Since the steady state error is zero the proportional gain was first designed.

The lead compensator was designed to increase the bandwidth. In this case the phase margin is enough to ensure stability. Gain was added instead using the equation:

$$Gain \ @ \ \omega_m = \sqrt{a_2} \qquad\qquad (4.36)$$

The gain added was $20dB$ at $300 \ rad/s$. And substituting results of equation and (4.33) into (4.31) the lead compensator was designed to be:

$$G_2 = \left(\frac{1.2s + 1}{0.02s + 1}\right)$$

To increase the speed a proportional gain was also designed for this loop. From , the characteristic equation of the voltage loop plant is:

$$sC_2 + K + Kr_{c2} + KsC_2 = 0$$

The pole position on the root locus to achieve the desired bandwidth and settling time is:

$$s = 2\pi(3501.42)$$

Substituting for "$s$" and the electrical elements:

$$K = 0.47781$$

The compensated closed loop is shown in Figure 4-11.

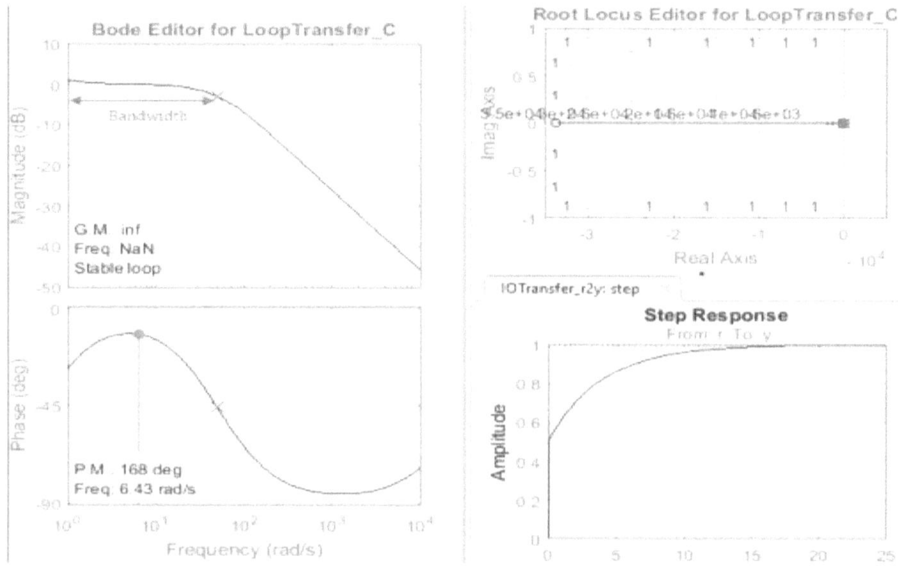


**Figure 4-11 Voltage loop lead compensator closed loop response**

The bandwidth of $314Hz$ now meets the specification. The output also shows no overshoot at the output. The phase margin has jumped to $168\ degrees$. The response now meets all the required specifications. The

# 5.  Bidirectional inverter simulated results

The inverter was simulated within MATLAB Simulink. The open loop results of the inverter DC-AC power flow were simulated analysing the output voltage and current. The inductor currents were then simulated followed by the voltage transfer ratio. The harmonic distortion was also analysed and then later the losses to determine the efficiency. The inverter AC-DC power flow was then taken through the same results analysis. Finally, the closed loop simulations were undertaken for both DC-AC and AC-DC power flows. All the results were simulated with dead time and parasitic elements – unless stated otherwise.

## 5.1    DC-AC power flow open loop simulations

The simulations were done with a fundamental fixed step size of $1 \times 10^{-4}s$ and using the solver: $ode23t(mod.\,stiff/Trapezoidal)$. The simulations were done from the Simulink models shown in Appendix A. All the simulations were done with a $60V$ DC input and $100\Omega$ AC load, unless stated otherwise.

### 5.1.1    *AC output voltage with a resistive load*

The duty cycle modulating signal used in the open loop was as analysed in Figure 3-10. Figure 5-1 shows the output voltage under ideal and non-ideal conditions. The inductor currents were simulated to show CICM operation and power flow direction.

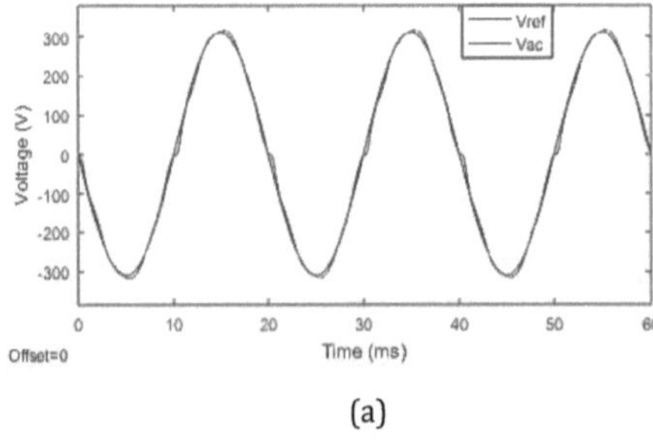


(a)

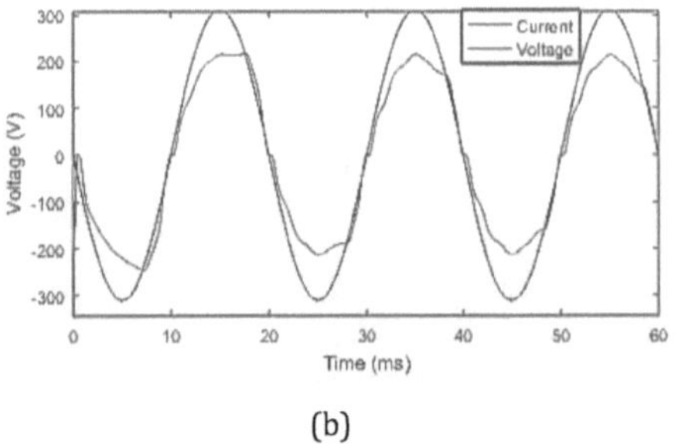


(b)

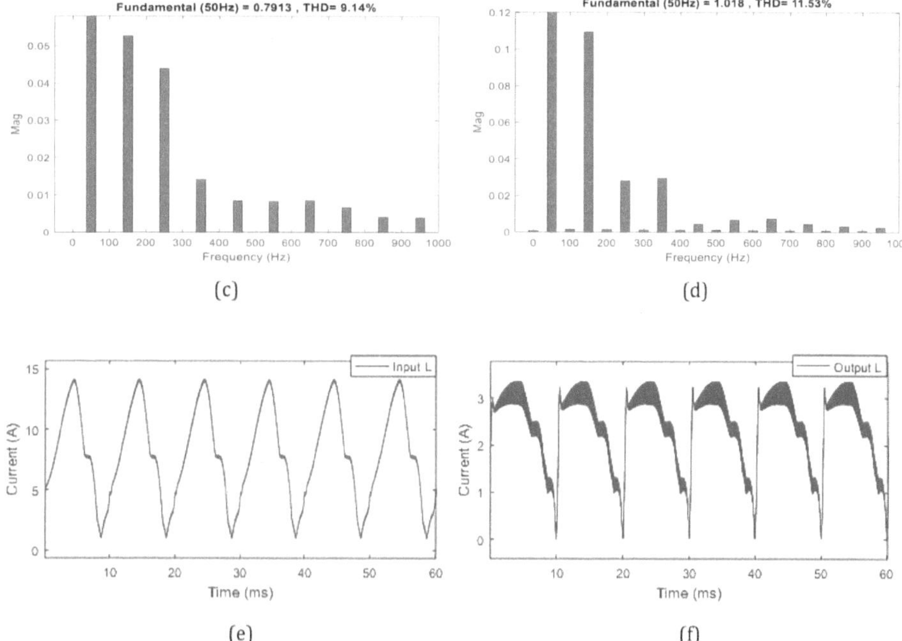

(c)         (d)

(e)         (f)

**Figure 5-1 Open loop results with restive load (a) output voltage without parasitic elements and dead-time (b) output voltage with parasitic elements and dead-time (c) FFT for output voltage with no dead time or parasitic elements (d) FFT for voltage with dead time and parasitic elements (e) input inductor current (f) output inductor current**

The output voltage is shown in Figure 5-1 (a) to be inverted as expected and it reaches the required $320.12V$ at 83.23% duty cycle but overshoots by approximately $9V$. It is also shown to be slightly out of phase with the reference signal. There is a steady state error of 12% of the terminal voltage. The total harmonic distortion of the voltage is 9.14% as shown in Figure 5-1 (c). The voltage also shows a small dead band at zero crossings because of the mismatch between the converter and unfolding stages. Figure 5-1 (b) shows that the output voltage is distorted. It doesn't reach the required peak voltage but rather reaches $200V$ peak. The THD shown in Figure 5-1 (d) has now increased to 11.53% which is above the desired specification. The steady state error it has is 24% of the peak value. The dead time and parasitic effects lower the RMS to $141V\ RMS$ and lower the maximum gain to 3.33. The output voltage also shows a small dead band at zero crossings as a result of the dead time in the unfolding bridge, as analysed in section 3.5.3, and the mismatch between the rectified voltage and the unfolding bridge. Figure 5-1 (e) shows that the peak current drawn from the supply is $15A$. The DC average of the current is $7.42A$ which is equivalent to the battery discharge current. Figure 5-1 (f) shows the output inductor with a large high frequency ripple than the input inductor current. This is because of the filter capacitor designed and the lower load current as analysed in section 4.3.3. Both inductor currents are continuous and positive which shows that they operate in CICM and the power flows from DC side to the AC side. The output power factor was 0.99, which is close to unity.

### 5.1.2    AC output voltage with an inductive load

The inductive load used was $50 + j15.70\Omega$. The output voltage and current were graphed on the same plot – using the code in Appendix B3 – to show the power factor. The left axis was the voltage and the right axis was the current. The inductor currents were also shown to indicate the power flow.

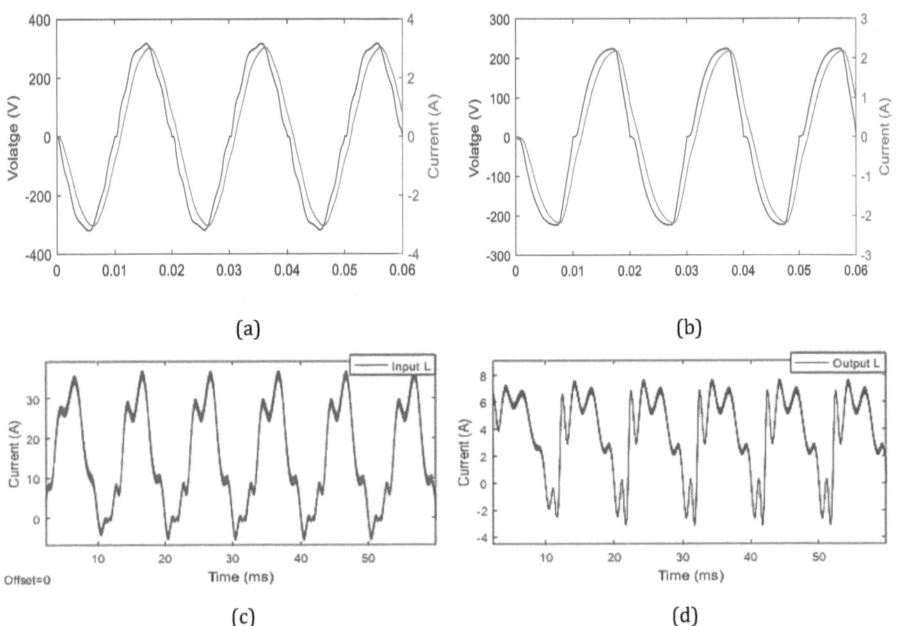

Figure 5-2 Open loop results with inductive load (a) output voltage without dead time and parasitic elements (b) output voltage and current with dead time and parasitic elements (c) input inductor current (d) output inductor current

Figure 5-2 (a) shows that the voltage has a peak voltage of $310V$ with a THD of 5.16% - an increase from the resistive load. Figure 5-2 (b) shows a voltage with peak of 187V a decrease from the resistive load. The lower values of voltage and higher THD values are because the inductance is frequency dependant – which distorts the voltage. Figure 5-2 (a) and (b) show the voltage and current are out of phase by $32^o$. Because the current is lagging, the power factor is $-0.85$. Figure 5-2 (c) and (d) show that the inductor currents reach negative values showing that the power is flowing from the AC side back to the DC side as analysed in section 3.2. This results in the four modes of operation as analysed in the same section.

## 5.2     AC-DC power flow open loop simulation

The AC-DC power flow simulations were also done with a Fundamental Fixed Step size of $1 \times 10^{-4}s$ and using the solver: $ode23t$ $(mod.\,stiff/Trapezoidal)$. The simulation was done from the circuit also in Appendix A1. As designed for in section 4.1, the reference charging voltage chosen for a $60V$ battery was $74V$. The nominal load resistor was $12.33\Omega$ - unless stated otherwise. This was chosen such that the battery charging current would be half the maximum charging current i.e. 6A. This is a charging rate of 0.5C which is the rate at which most batteries are charged. All the simulations were done with dead time, parasitic elements and the source inductance included – unless stated otherwise.

### 5.2.1     Output DC voltage

As per the specifications the input voltage used was $\sqrt{2}(220)\sin(2\pi \times 50t)$ and the load resistance was $12.33\Omega$. The charging voltage reference is $74V$. This reference was filtered such that its settling time was 30ms as per the design specifications. The duty cycle modulating signal used was as analysed in Figure 3-16 during the analysis. Figure 5-3 shows 2 DC output voltage outputs each with a different set of conditions to assess the effects of the AC source inductance and dead time and parasitic elements on the output charging voltage.

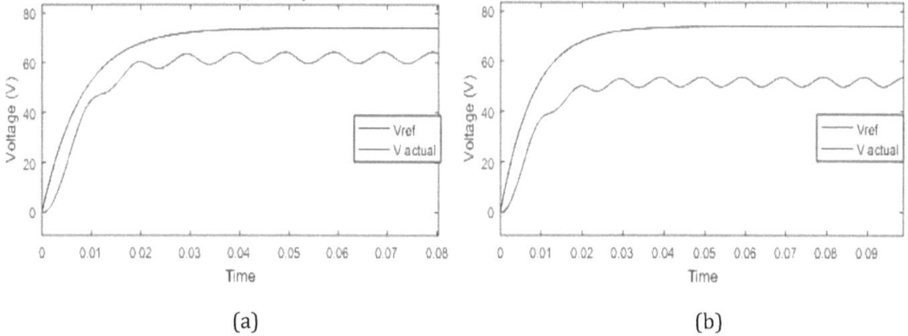

(a)                                                       (b)

**Figure 5-3 AC-DC power flow DC output voltage results a) source inductance included but no dead-time and parasitic elements b) with parasitic elements and dead time and no source inductance**

In Figure 5-3 (a) the output voltage is shown to settle in $24.89ms$. This is $15ms$ faster than the required settling time. It has no overshoot which is desired. It also shows large oscillations at the output which because of the varying duty cycle. The steady state error observed is $13V$ which would be insufficient to charge the battery. Figure 5-3 (b) shows that the output voltage has lowered to a final value of $50V$ with oscillations that are lower than observed in (a); this is because the parasitic elements and dead time affect the gain more than the source inductance as analysed. The settling time was $24.67ms$ which is within the specifications.

### 5.2.2 Input coupling capacitor voltage

The Ćuk inverter's AC-DC power flow operates in DCVM where input capacitor voltage is discontinuous. The voltage appearing across the capacitor was observed.

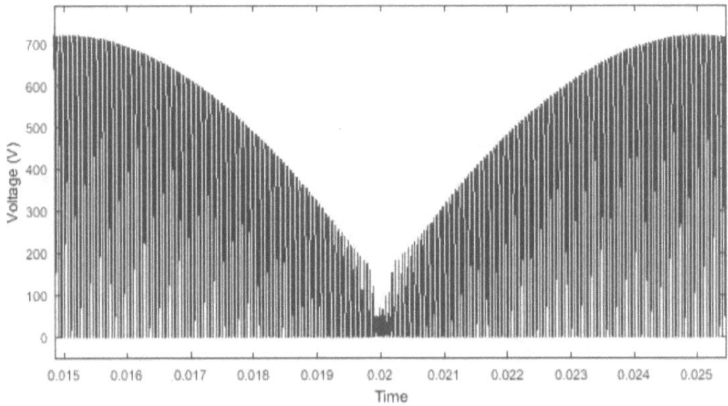

**Figure 5-4 AC-DC power flow input coupling capacitor discontinuous voltage**

Figure 5-4 shows that the capacitor voltage is discontinuous as designed for. The waveform shows that at every half cycle, there is a continuity in the voltage at the zero crossing, this is because the input voltage switches from positive to negative and the rate at which the inductor charges and discharges is higher than that of the fundamental frequency.

### 5.2.3 Input power factor

The Bidirectional inverter was designed to work in DCVM during AC-DC power flow to ensure that the Input power factor (PF) is as close to unity as possible to ensure high efficiency and low EMI. Having observed that the circuit does operate in DCVM, the PF at the AC side during load regulation was assessed. In Figure 5-5 (b) the load current was varied from $3 - 12A$.

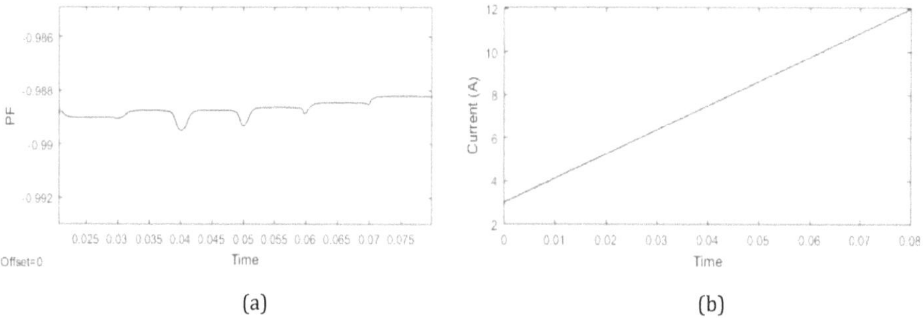

(a)          (b)

**Figure 5-5 AC-DC input power factor during load regulation (a) power factor (b) load current**

The power factor in Figure 5-5 (a) is shown to decrease as the load current shown in Figure 5-5 (b) increases. It decreases by an order of 1000 but still above $-0.98$ within the operating range of the inverter. The PF decreases because the system is put under more strain and the line current distorts which leads to the lower PF.

## 5.3 DC-AC power flow inverter closed loop simulation results

The closed loop was simulated to assess the effectiveness of the nested loop control strategy designed. It was then put through a series of disturbance rejections to assess its ability to handle variations. All the closed loop simulations were done with dead time and parasitic elements included.

### 5.3.1 AC output voltage with a resistive load

The same conditions as the open loop simulations were kept and the closed loop simulated. The input voltage was kept constant at $60V$ and load resistor was kept at $100\Omega$. Figure 5-6 shows the output voltage.

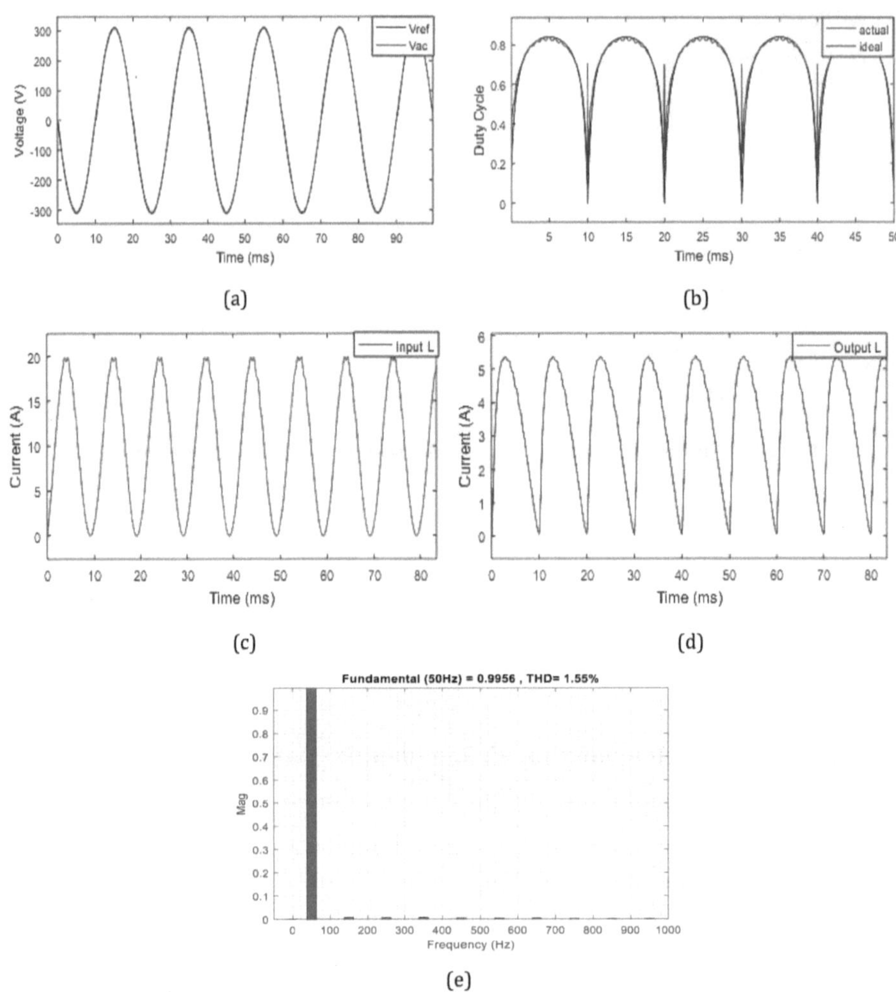

**Figure 5-6 Closed loop DC-AC power flow resistive load results (a) AC output voltage (b) duty cycle modulating signal (c) input inductor current (d) output inductor current (e) AC output voltage FFT**

Figure 5-6 (a) shows that the AC voltage has improved from the open loop system with dead time and parasitic elements. The RMS voltage observed is $219.94V$ and is within the required regulation specification. The THD shown in Figure 5-6 (e) is now 1.55% which is within the required specification. The THD also meets all the required grid harmonic current specification as reviewed in Table 2-3 – Table 2-5 for various residential loads. Figure 5-6 (b) shows that the duty cycle modulating signal is shown to follow the ideal profile. It shows small oscillations when the duty cycle reaches its peak – this is because of the parasitic elements within the circuit. There are also duty cycle spikes observed at half cycles because of the control action compensating for the mismatch between the converter stage and unfolding bridge stage and the dead time during the unfolding stage. The inductor currents shown in Figure 5-6 (c) and (d) have improved to show less distortion. The average current drawn from the supply is now $10A$. The output power factor is unity as expected from a resistive load.

### 5.3.2    *AC output voltage with an inductive load*

The same inductive load as the open loop system was used to assess the power factor and power flow with an inductive load. The AC voltage and current were again plotted on the same graph to illustrate the phase difference between voltage and current with an inductive load.

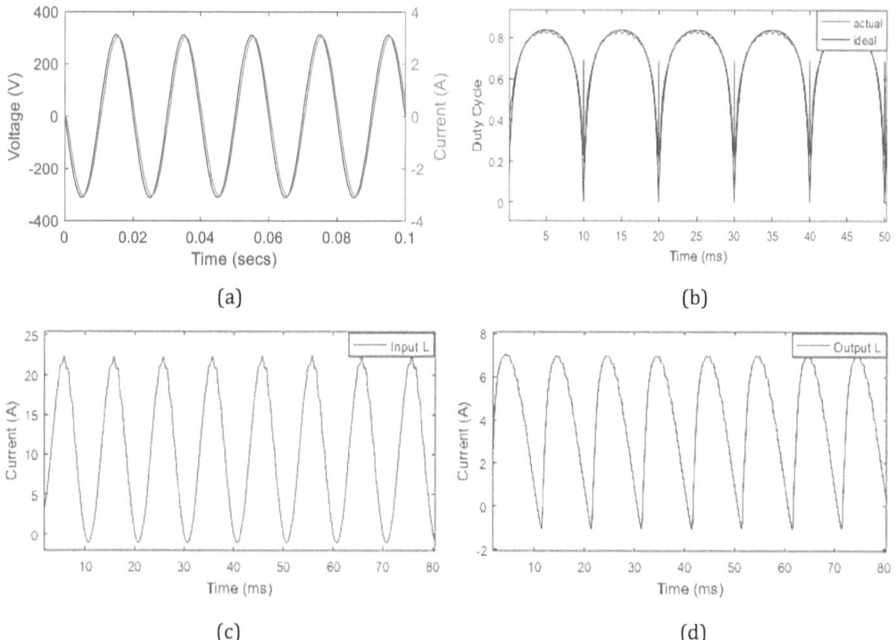

**Figure 5-7 Closed loop DC-AC power flow inductive load results (a) AC voltage (b) duty cycle modulating signal (c) input inductor current (d) output inductor current**

Figure 5-7 (a) shows that the phase difference between the voltage and current has improved from $32^o$ in the open loop, to $11^o$ in the closed loop – this is power factor of 0.98. The duty cycle shown in Figure

5-7 (b) is the same as the resistive load ensuring that the voltage RMS is 219.84$V$ – a 0.1$V$ difference from the resistive load.

### 5.3.3    *Closed loop AC load and line regulations*

In Figure 5-8 (a), the input voltage was kept at a constant 60$V$. The load current was then set to a maximum and the output voltage noted; the load current was then set to a minimum and the output voltage noted again. The output current was varied continuously from 0.91$A$ *RMS to* 4.54$A$ *RMS*, as per design specification, and equation (3.58) used to simulate the load regulation. In Figure 5-8 (b) the load current was kept constant at 2.12$A$ *RMS.* The input voltage was then set to a maximum and the output voltage noted; the input voltage was then set to a minimum and the output voltage noted again. The input voltage was then varied from $60 - 70V$ and using equation (3.59), the line regulation was simulated.

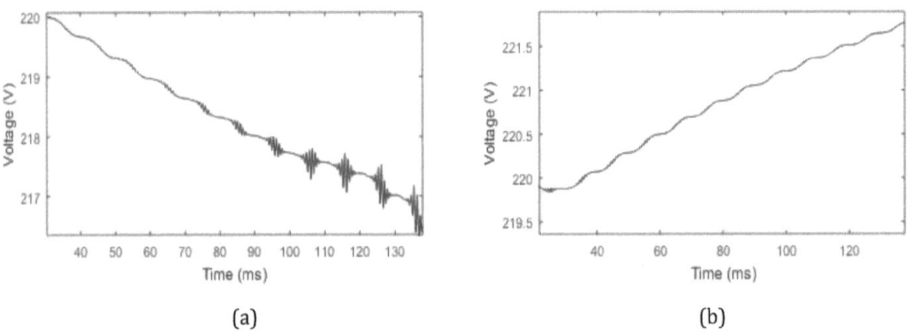

(a)          (b)

**Figure 5-8 Closed loop regulations (a) load (b) line**

The load regulation in Figure 5-8 (a) is now shown to from $-0.03\%$ to $-1.8\%$ – which is within the required regulation range. The line regulation in Figure 5-8 (b) is from $-0.03\%$ to $0.82\%$ – which is also within the required regulation range. The AC voltage fluctuation is less when the DC voltage is varying than when the load is varying, this is because the load regulation range is wider the line regulation range; and increasing the input voltage means that the control action isn't strained and hence, the duty cycle modulating signal produced by the control, is better under line regulation. This also explains the ripples observed in Figure 5-8 (a) at higher load current values.

### 5.3.4  *RMS voltage set point tracking*

To check for the tracking specification, the set point RMS voltage was stepped down from $220V$ $RMS$ $to$ $120V$ $RMS$ at 50ms to determine the tracking ability of the control strategy.

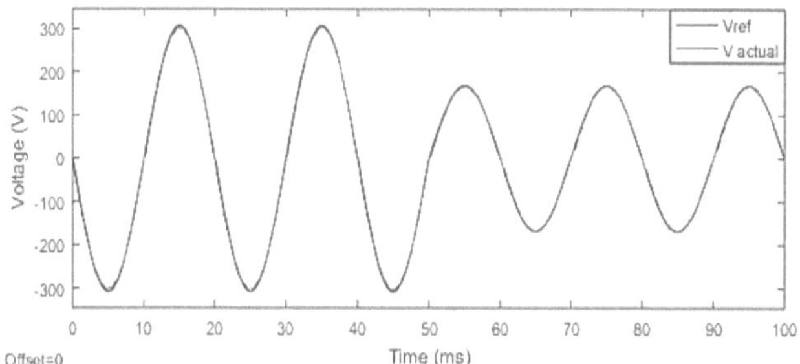



**Figure 5-9 DC-AC power flow set point 220V to 120V RMS tracking**

Figure 5-9 shows that the output voltage references are tracked with no steady state error. After $50ms$ the peak shows to have dropped from $220V$ to $120V$.

**Table 5-1 RMS Reference tracking output voltage performance comparison between 120V RMS and 220V RMS**

| Test Variable | Ideal RMS Value ($V$) | Peak Value ($V$) | RMS ($V$) | %THD |
|---|---|---|---|---|
| **120$V$ RMS reference** | 120 | 169.91 | 119.97 | 1.28 |
| **220$V$ RMS reference** | 220 | 310.75 | 219.94 | 1.55 |

Table 5-1 shows that the system better tracks the $120V$ $RMS$ voltage reference. This is because the control action doesn't work as hard, since the duty cycle required to step to a lower RMS is lower and hence, the parasitic effects are less.

### 5.3.5  *Input and output disturbance rejection test*

The Inverter was taken through some disturbance rejection tests. The nominal input voltage used was $60V$ and nominal load resistor used was $100\Omega$ or $3A$ load current unless stated otherwise. The first 2 were a load disturbance: a load addition and a removal; the next was inrush current; and the last was a non-linear load test.

*i.*      **Input voltage step**

The input voltage disturbance rejection was assessed by stepping the input voltage from $60V$ to $65V$ at $50ms$; it was then stepped from $65V$ to $70V$ at $100ms$. This is to check how the system responds to sudden changes in the input voltage.

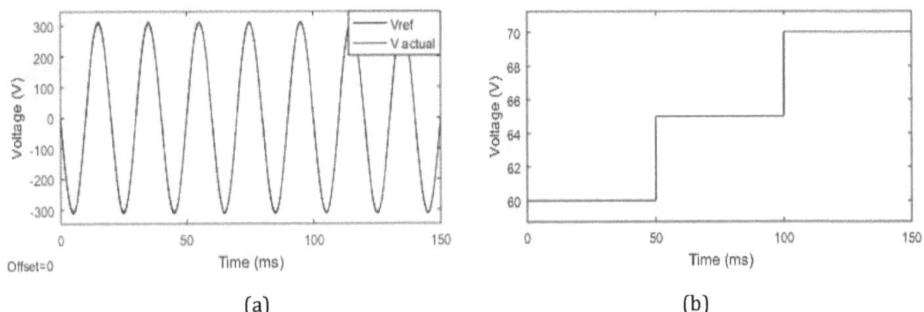

(a)                     (b)

**Figure 5-10 Input voltage disturbance rejection (a) AC output voltage (b) DC input voltage**

As the input voltage is stepped, the output voltage remains constant as shown in Figure 5-10 (a). The harmonic distortion decreases as the voltage increases. The controller can handle input disturbances from the input. Table 5-2 shows these comparisons.

**Table 5-2 Input voltage disturbance output voltage performance comparison**

| Test Variable | Peak Value ($V$) | RMS ($V$) | %THD |
|---|---|---|---|
| **Ideal Voltage** | 311.12 | 220 | 0 |
| **60V** | 310.85 | 219.94 | 1.55 |
| **65V** | 310.96 | 219.97 | 1.18 |
| **70V** | 311.19 | 220.14 | 1.04 |

The output voltage performance improves as the input voltage increases. This is because the control action does not work as hard when the input voltage is high. The duty cycle required to increase the peak decreases which lessens the effects of parasitic elements within the inverter.

*ii.*       ***Load removal***

In residential system being designed for, a sudden decrease in the load current results from switching off lights or other equipment, within the critical loads panel. A 200Ω resistor was added at $50ms$ of the simulation to decrease the load current and observe the system response.

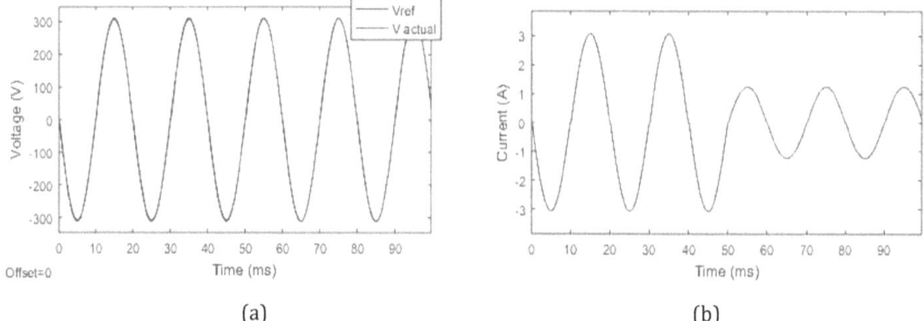


(a)                                    (b)

**Figure 5-11 Load removal disturbance rejection test a) voltage b) current**

Figure 5-11 (a) shows that the voltage does not change i.e. the control loop rejects the load addition and ensures that the output voltage is at the required peak. Figure 5-11 (b) shows the output current decreases from $3A$ to $1A$ peak. Table 5-3 shows the comparisons.

**Table 5-3 Load disturbance output voltage performance comparison between 100 and 200Ω load resistors**

| Test Variable | Peak Value ($V$) | RMS ($V$) | %THD |
|---|---|---|---|
| **Ideal Voltage** | 311.12 | 220 | 0 |
| **2.12$A$ Load** | 310.85 | 219.94 | 1.55 |
| **0.9$A$ Load** | 311.19 | 220.04 | 1.12 |

The lighter load performs better than the heavier load. This is largely due to the fact that in section 3.6.1 it was shown that the larger the load resistor compared to the parasitic inductor resistors, the higher the gain attained by the inverter.

*iii.* ***Load addition***

A 50Ω resistor was removed from the output at $50ms$ to increase the load to rated current and the output voltage and current were shown.

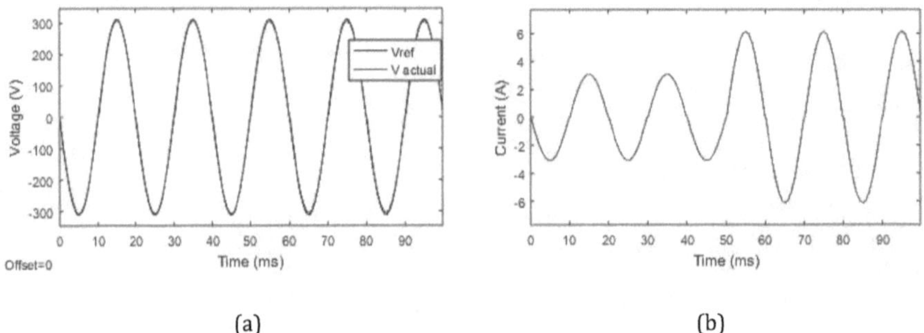

(a) (b)

**Figure 5-12 Load addition disturbance rejection a) voltage b) current**

Although the voltage in Figure 5-12 (a) rejects the load current sudden change, it is shown to have a larger steady state error and ripples at the peak values when operating at rated current. The current in Figure 5-12 (b) increases accordingly due to less resistance at the load output. The current is small ripples at rated AC current.

**Table 5-4 Load disturbance output voltage performance comparison between a 100 and 50-ohms load resistor**

| Test Variable | Peak Value (*V*) | RMS (*V*) | %THD |
|---|---|---|---|
| **Ideal Voltage** | 311.12 | 220 | 0 |
| **2. 12$A$ Load** | 310.85 | 219.94 | 1.55 |
| **4. 45$A$ Load** | 310.63 | 219.73 | 2.19 |

Table 5-4 shows that the inverter performs better under the lighter load as discussed in the previous sub-section. Although at it doesn't perform as well at rated current, it still meets the required specification.

*iv.* ***Inrush current***

The system was tested for handling of inrush currents at $50ms$ in Figure 5-13. The system used to simulate inrush currents was an H-bridge diode with a RC load of $100 - j25.16Ω$. This is because the H-bridge has a square wave transfer function as shown in the analysis. As such, the current jumps from a low to high value quickly. The sum of the inductor currents was limited to a maximum of $70A$ and a minimum of $-20A$ in the control saturation block.

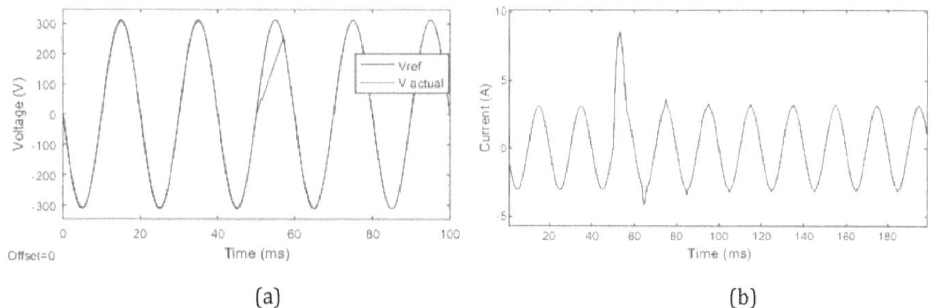

Figure 5-13 Inrush current disturbance test (a) voltage (b) current

Figure 5-13 (a) shows that the voltage deforms at $50ms$ when the inrush current is introduced. The controller ensres that the voltage recovers within half a period as shown. The current in Figure 5-13 (b) is shown to have a large spike of $8A$ at $50ms$ but recovers also in less than half the period. The controller shows that it can handle inrush currents.

*v.*     ***Non-linear load***

A non-linear load in Figure 5-14 was added at the output instead of the linear resistor to determine the ability of the controller to provide regulated RMS line output voltage to a non-linear load such as TV. The non-linear load was modelled using a full bridge rectifier with a shunt $400\Omega$ load resistor.

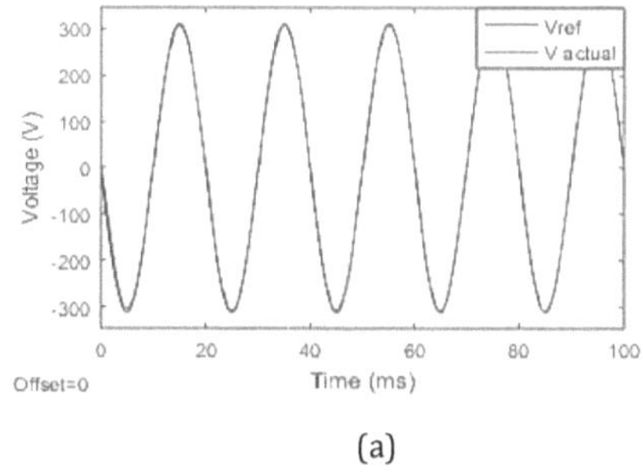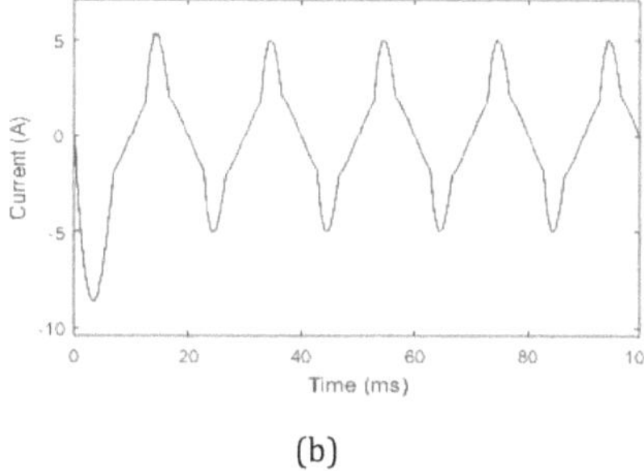

(a)             (b)

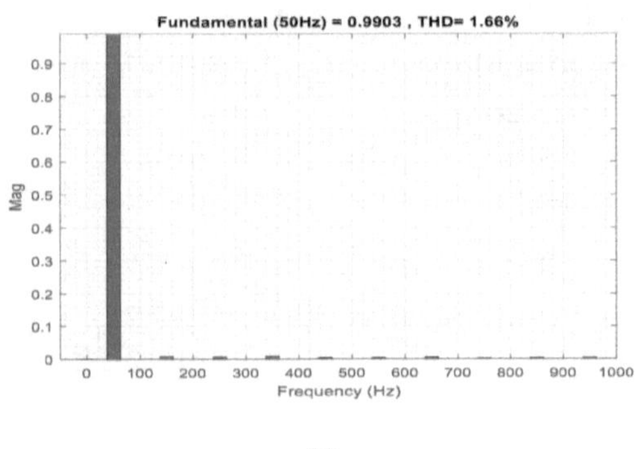

(c)

**Figure 5-14 Non-linear load disturbance test (a) output voltage (b) output current (c) output voltage FFT**

Figure 5-14 (a) shows that the output voltage still provides the required line voltage with minimal steady state error and harmonic distortion. The settling time for the load output voltage also increased to $15ms$; this is still within the required specification. The THD shown in Figure 5-14 (c) has increased to 1.66% from the linear load. and the steady state error is $1.12V$. The current in Figure 5-14 (b) shows the non-linearity and a peak amplitude of $5A$.

## 5.4    AC-DC power flow closed loop simulation results

The closed loop was designed to solve the short-comings of the open-loop. Like the DC-AC power flow, the AC-DC power flow was also subjected to various input and output disturbances. All the closed loop simulations were done with dead time, parasitic elements and the AC source inductance included unless stated otherwise.

### 5.4.1    Line voltage and current

This simulation shows reverse power flow capability of the bidirectional inverter. The line voltage of $200V\ RMS$ was scaled down by a factor of a 100 – since it is larger than the line current – to clearly show the phase difference between the two signals. Dead time and parasitic elements and the AC source inductance were neglected. The inductor currents were also simulated to show the direction of power flow.

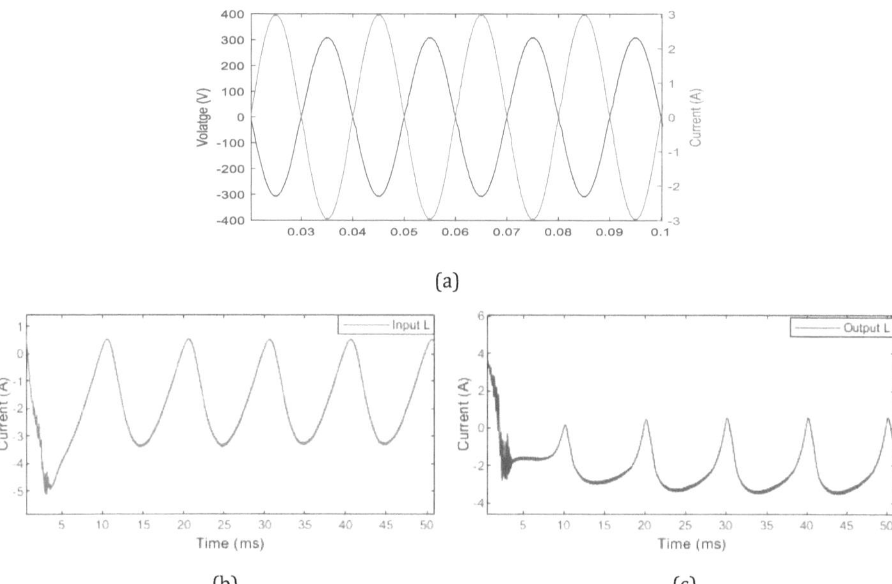

(a)

(b)                                    (c)

**Figure 5-15 AC-DC power flow (a) line voltage and current (b) input inductor current (c) output inductor current**

Figure 5-15 (a) shows that the line current is distorted near the peaks. line voltage and current are out of phase by approximately $562.03\mu s$ which is a power factor of approximately $-0.9844$. This shows that the power is currently flowing from the AC source back to the battery. Of all the power delivered by the AC source, only $0.00001W$ is reactive power. Figure 5-15 (b) and (c) shows that the inductor currents are negative indicating that the power flows from the AC side back to the battery.

### 5.4.2 *Output DC voltage*

The plots in Figure 5-16 were tested at a $12.33\Omega$ load resistor and $220V$ *RMS* AC line voltage.

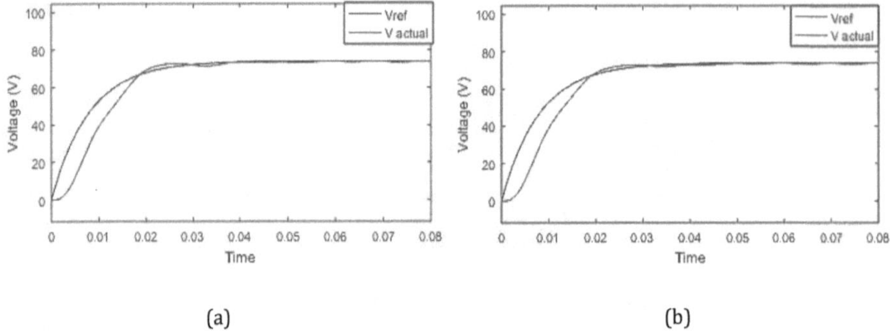

(a)            (b)

**Figure 5-16 AC-DC power flow closed loop output voltage (a) with source inductance and not dead time and parasitic elements (b) with dead time and parasitic elements and no source inductance**

The output voltage in Figure 5-16 (a) is shown to track the reference voltage with a $0.24V$ steady state error. The settling time is $32.89ms$ which is within the desired specification. The oscillations at the output have been reduced. The output voltage in Figure 5-16 (b) settles faster at $29.98ms$ with a steady state error of $0.35V$. The voltage shows a slight increase in oscillations. The control therefore ensures a constant regulated charging voltage for the battery.

### *5.4.3    Output voltage set point tracking*

The inverter was also put through some set point changes to check its ability to track various steps, $(64V, 74V, 84V)$.

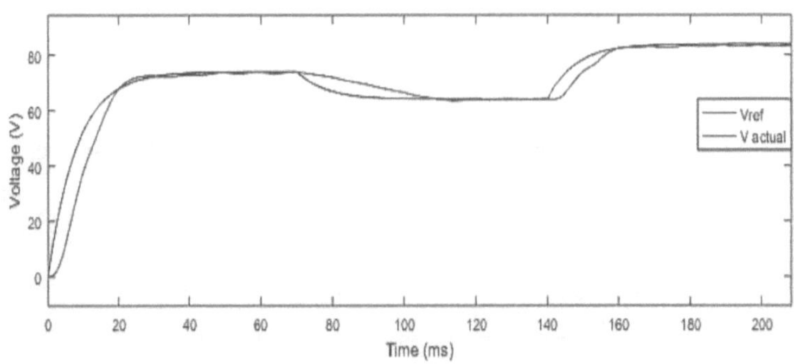

**Figure 5-17 Closed loop voltage set point tracking**

The output voltage waveform set point in Figure 5-17 tracks the set points. For the first $70ms$, a $74V$ set point is tracked with little steady state error and no overshoot. The 64V step is shown to settle the slowest and the $84V$ step settles the fastest. The ripple for the $84V$ step is more significant than the other steps. Table 5-5 shows the performance comparison between the step changes.

Table 5-5 Voltage step tracking output voltage performance comparison

| Step references | Steady state error ($V$) | Settling time ($ms$) | % Ripple |
|---|---|---|---|
| **64V** | 0.21 | 45.22 | 0.9 |
| **74V** | 0.25 | 32.98 | 1.2 |
| **84V** | 0.56 | 29.54 | 2.26 |

The 64V step has the lowest steady state error and slowest settling time but the ripple at the output is smallest with this step. The $84V$ step has the largest steady state error and fastest settling time. This is because a smaller input voltage is used for a higher reference; this increases the duty cycle required to a point where the effect of parasitic elements is largest.

### 5.4.4    Disturbance rejections tests

There were two disturbance tests that were done. The first was a load disturbance test based on stages of charging a battery and the last was an inrush current test.

#### i.        DC load current disturbance rejection assessment

In Figure 5-18 (b) the load current was stepped from $3A$ to $6A$ charging current and then decayed back to $3A$ to observe the charging voltage when the charging current of the battery changes.

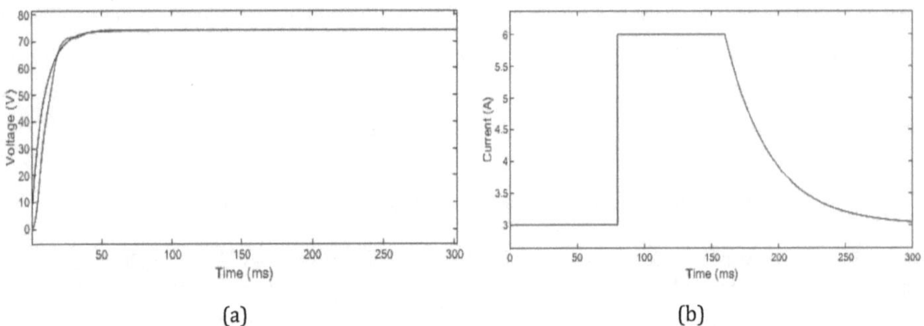

(a)                                        (b)

**Figure 5-18 DC load charging current disturbance rejection (a) charging voltage (b) charging current**

The DC voltage in Figure 5-18 (a) is also shown to have improved from the open loop. The charging voltage is now constant regardless of the charging current used. Table 5-6 shows the performance comparison between the different charging stages.

**Table 5-6 Charging stages simulation performance comparison**

|  | Steady state error ($V$) | Settling time ($ms$) | %Ripple |
|---|---|---|---|
| **Pre-charge** | 0.09 | 42 | 0.1 |
| **Constant Current** | 0.24 | - | 1.2 |
| **Constant Voltage** | 0.14 | - | 0.6 |

The inverter is shown to perform the best during the pre-charge stage because of the lighter load. All the stages met the required performance specifications.

### ii.      *Inrush current disturbance test*

So far, the disturbance tests were done during steady state but, when AC is rectified back into DC power, there are some inrush currents at start-up if the AC voltage is applied suddenly using contacts – this leads to a large surge power at the input. This test investigates the ability of the inverter and control to compensate for these inrush currents. The initial condition of the filer capacitor was set to zero.

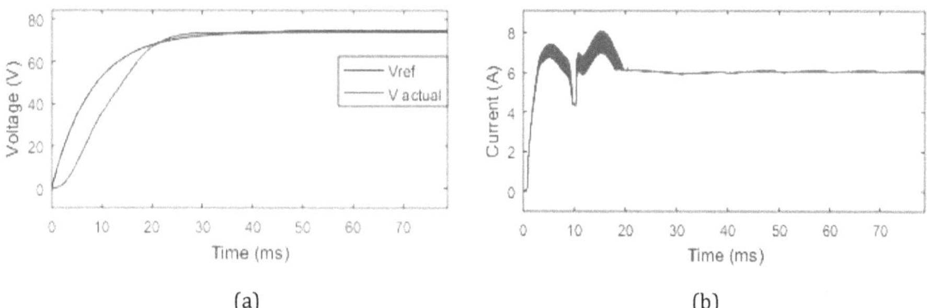

**Figure 5-19 Inrush current disturbance rejection (a) output voltage (b) output current**

The output voltage in Figure 5-19 (a) is shown to reject and changes in the response. It still tracks the output voltage reference with a steady state error of $0.85V$. The settling time has increased to $36.54ms$ because of the inrushing current. The output current in Figure 5-19 (b) is shown to have an overshoot of 33.3% of the final current value which settles in $20ms$. The current overshoot is above the required specification but within the maximum charging current of $12A$ for the battery.

# 6. Bi-directional inverter experimental results

The bidirectional inverter was built in the experimental and the practical results were detailed below. The experimental testing was made as close as possible to the simulated testing.

## 6.1 Experimental testing overview

The experimental set up and test procedure was detailed here. The set up was described in terms of a block diagram.

### 6.1.1 Experimental set-up

The Ćuk inverter system shown in Figure 6-1 was built on a Vero-board for easy manipulation. A DSPACE CP1104 digital signal processing unit was used. The unit has both digital and analogue input and output (I/O) pins.

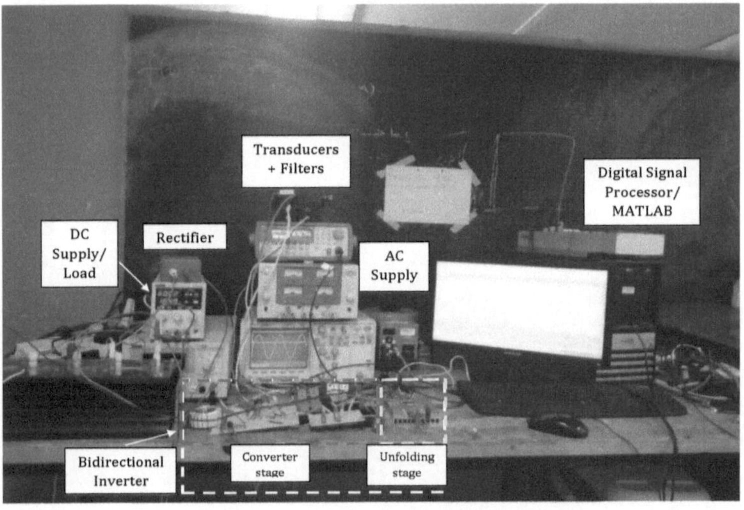


**Figure 6-1 Experimental set up**

The block diagram for the experimental set up Figure 6-1 is shown in Figure 6-2.

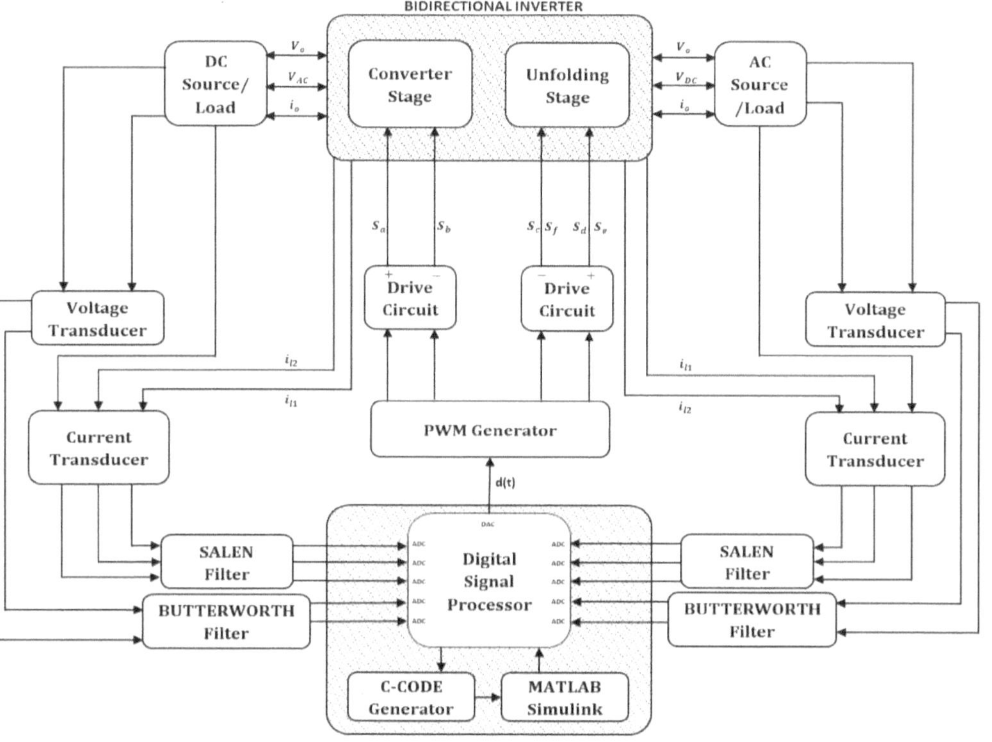


Figure 6-2 Experimental set up block diagram

It was interfaced with MATLAB Simulink through the Real-Time Interface. The Ćuk inverter output signals are converted into analogue signals and inputted into the DSPACE unit and then viewed on MATLAB Simulink. Simulink produces the duty cycle modulating signal which is the compared to high frequency oscillator carrier signal to produce the required pulses for the switches as shown.

### 6.1.2  *Inverter testing strategy*

The Bidirectional inverter was tested under the consideration of 3 controlled variables. These variables are the only variables that can be changed at the discretion of the designer to emulate the real-world conditions. As they vary, they affect the performance of the inverter – as analysed in 3.4 – They were discussed in more detail.

### i.     *Load power*

The load power was controlled by adjusting the DC or AC load current for the bidirectional inverter. As it varied, it affected the output voltage, %THD and efficiency of the inverter. This is because in real systems the load power is not constant all the time. Based on the demand in a typical residential AC load, the output power delivered by the inverter will also vary.

### ii.     *DC line voltage*

The line voltage can be controlled easily by varying the voltage source. In real systems, the Line voltage fluctuates from a maximum to a minimum depending on the conditions that affect the energy source such as age and ambient temperature where it's operating. This fluctuation will affect the efficiency, power output and %THD of inverter.

### iii.     *Duty cycle*

The maximum duty cycle also varies depending on what the requirements are, to keep the load regulated. As it varies, it also affects the gain, output voltage and mostly efficiency due to the poor switch utilisation at higher duty cycles.

## 6.2     Switching scheme results

The bidirectional inverter was designed such that the switches operate using zero-voltage switching to reduce the switching losses and so that during AC-DC power flow discontinuous capacitor voltage mode can be achieved.

### 6.2.1     *IGBT switching signals*

The switching signals for the IGBT with dead time included are shown in Figure 6-3.

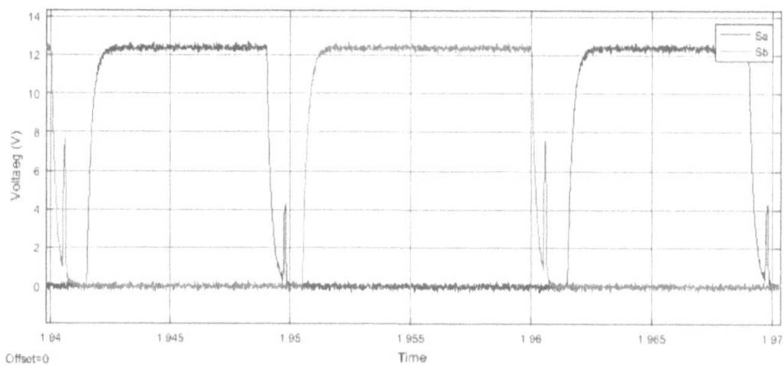

Figure 6-3 Ćuk-Converter switching signals with dead time inserted

The switching signals are shown to be conjugates of one another. They show that they are never both ON at the same time due to the dead time. The shown resonant spikes which are small and only last for a short time. Also, they do not affect the switches drain to source voltage. These spikes are caused by the dead time.

### 6.2.2    Converter stage IGBT drain-source voltage and current

The IGBT converter switch's drain to source voltage and current were tested to observe the designed switching scheme.

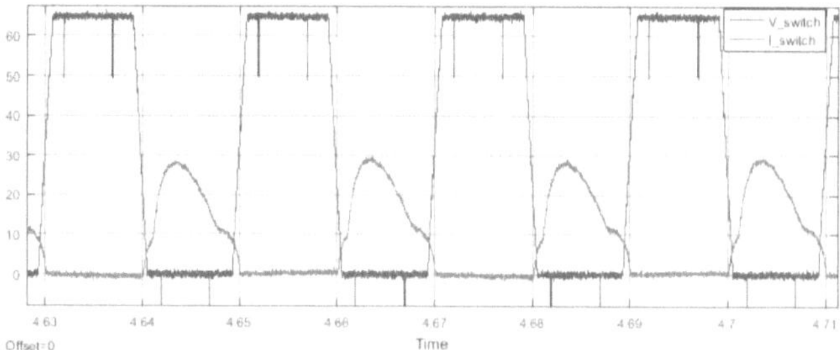

Figure 6-4 Bi-directional inverter switching voltage and current waveforms

Figure 6-4shows the switch voltage at a peak of $65V$ and the switch current at a peak of $27.89A$, it shows that the zero-voltage switching was not achieved. There is still an overlap between the switch voltage and switch current during the switch's ON-OFF region. The figures also show a small leakage current of $13mA$

during switch ON. The ON-state voltage is shown to be $0.09V$. Table 6-1 shows rise and fall times of the converter stage IGBTs.

Table 6-1 Converter stage IGBT switching time characteristics

|  | $t_{rise}(ns)$ | $t_{fall}(ns)$ |
|---|---|---|
| **Drain to source voltage** | 116.33 | 95.68 |
| **Drain to source current** | 184.05 | 249.72 |

These times under achieved ZVS would be zero. The ON-OFF State losses are given by the voltage rise time and the current fall time. The OFF-ON State losses are given by the voltage fall time and the current rise time. Table 6-2 shows the Unfolding bridge switching time characteristics.

Table 6-2 Unfolding bridge IGBT switching time characteristics

|  | $t_{rise}(ns)$ | $t_{fall}(ns)$ |
|---|---|---|
| **Drain to source voltage** | 378.15 | 239.38 |
| **Drain to source current** | 539.75 | 758.34 |

The rise and fall overlap times for the Unfolding bridge IGBTs are shown to be higher than those during the converter stage IGBTs. This is because they are switched at a lower frequency, but they dissipate less power since the output current is much lower than the input current since the inverter is steeping up and the fundamental Grid frequency is also lower than the carrier frequency.

### 6.2.3    IGBT switching losses

The losses associated with the switch were tabulated and calculated using the parameters discussed in the previous section, Table 6-1 and Table 6-2; and the equations as analysed in 3.7.1i – the losses were calculated.

Using (3.120), the losses during the period OFF-ON of the switch are:

$$P_{c(on)} = 167.40mW \text{ for } S_a \text{ and } S_b; \text{ and } P_{c(on)} = 4.69mW \text{ for } S_c, S_f, S_e, S_d$$

Using (3.121), the losses during the period ON-OFF of the switch are:

$$P_{c(off)} = 237.93mW \text{ for } S_a \text{ and } S_b; \text{ and } P_{c(off)} = 9.31mW \text{ for } S_c, S_f, S_e, S_d$$

Using (3.122), the ON-State losses because of the switch voltage not reaching zero but reaching a minimum of 0.09V during the ON time on the switch are:

$$P_{on} = 7.94W \text{ for } S_a \text{ and } S_b; \text{ and } P_{on} = 3.21W \text{ for } S_c, S_f, S_e, S_d$$

Using (3.123), the losses due the leakage current during switch OFF are:

$$P_{off} = 845mW \text{ for } S_a \text{ and } S_b; \text{ and } P_{off} = 589mW \text{ for } S_c, S_f, S_e, S_d$$

The losses shown above account for approximately $28W$ of power loss. This power-loss is mainly due to the ON-State voltage of the switch which was not at zero volts during switch ON.

## 6.3     DC-AC power flow inverter open loop experimental results

All the experimental results include dead time and parasitic elements due to the practical elements being used. Instead of batteries as designed for, a DC power supply used due to a shortage of batteries. Only the resistive and inductive load testing was done during the open loop because the simulations already showed that the open loop regulation tests were poor – this was to avoid redundancy. The inductor currents were obtained from the LEM modules and viewed in Simulink through the DSPACE unit.

### 6.3.1     *AC output voltage with resistive load*

The output voltage and current were tested for using an input voltage of $65V$ because an input voltage of $60V$ produced less than adequate results during the preliminary testing because of the high DC resistance of the wires and inductors; and a load resistor of $100\Omega$. The input side included a coupling capacitor of $3.1mF$ due to the unregulated input current during the experimental testing.

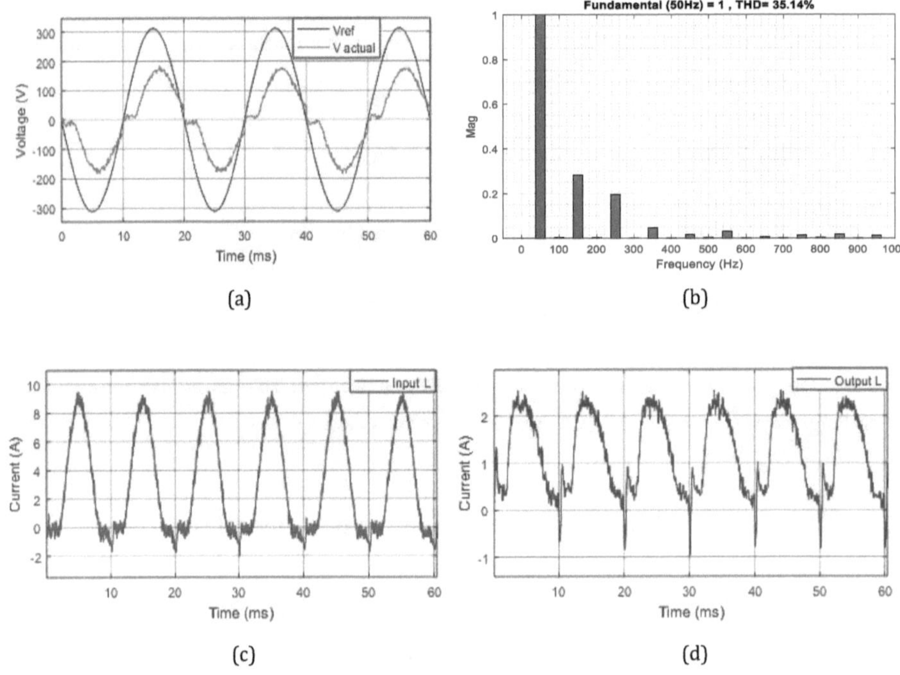

(a)

(b)

(c)

(d)

**Figure 6-5 DC-AC inverter output experimental results (a) voltage (b) current (c) input inductor current (d) output inductor current**

Figure 6-5 (a) and (b) show that the output voltage and current are both sinusoidal but are deformed and have noise. The output voltage reaches a peak of $187.23V$ instead of $311.13V$ – an RMS value of $132V$. The maximum gain attained by the open loop system is 2.87. The THD shown in Figure 6-5 (b) is 35.14% - which is above the desired specification. This distortion is caused by the dead bands at zero crossings. These dead bands are more significant than those during the simulations. They are caused by the dead time during the unfolding stage and the mismatch between the converter and unfolding stages. This dead band is approximately $2.5ms$. Figure 6-5 (c) and (d) show that both input and output inductor currents are continuous as designed for. The input inductor shows the lower current drawn from the supply. The power is still flowing from the DC supply to the AC load. Although there is a high distortion in the voltage and current waveforms, the power factor is still 0.98.

## 6.3.2     AC output voltage with an inductive load

An inductive load of  $50.35 + j19.92\Omega$  was used. This was close to the inductive load used during the simulations but also included the inductor ESR. The waveforms were then observed.

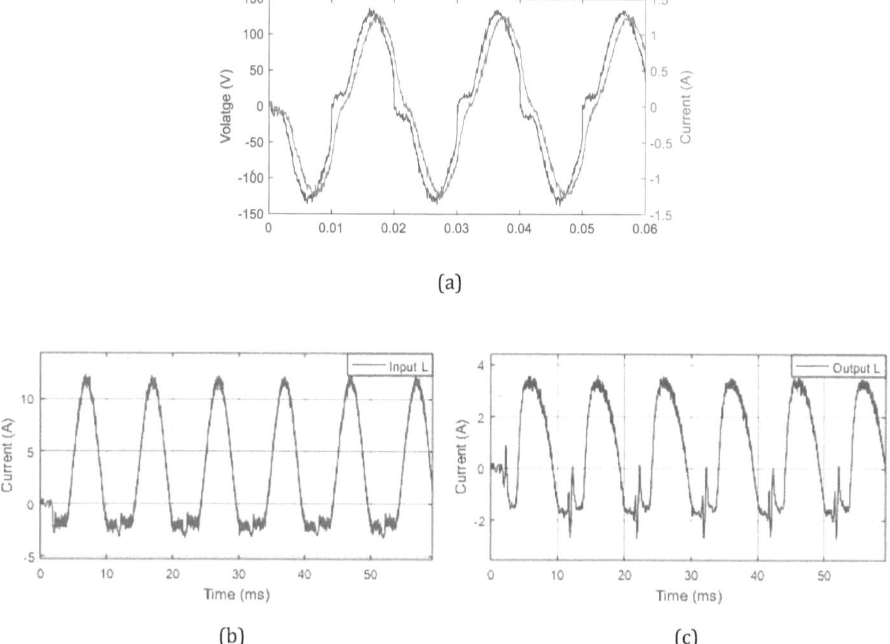

(a)

(b)          (c)

**Figure 6-6 DC-AC power flow inductive load results (a) voltage and current (b) input inductor current (c) output inductor current**

Figure 6-6 (a) shows the voltage and current with a phase difference of $35^{o}$ – a lagging power factor of $-0.82$. This slightly lower than the simulated results. The inductor currents in Figure 6-6 (b) and (c) shows the reverse power flow.

## 6.4 AC-DC power flow experimental results

The reverse power flow was tested and inherently, the source inductance and parasitic elements could not be ignored. The output voltage and current were tested for first and then the coupling capacitor voltage was assessed.

### *6.4.1 Output voltage*

An AC power supply was used as an input source and a DC resistive load of 12.33Ω to emulate the battery's behaviour when it charges at 0.5C which gives a charging current of 6A. The RMS input voltage was set to 220V.

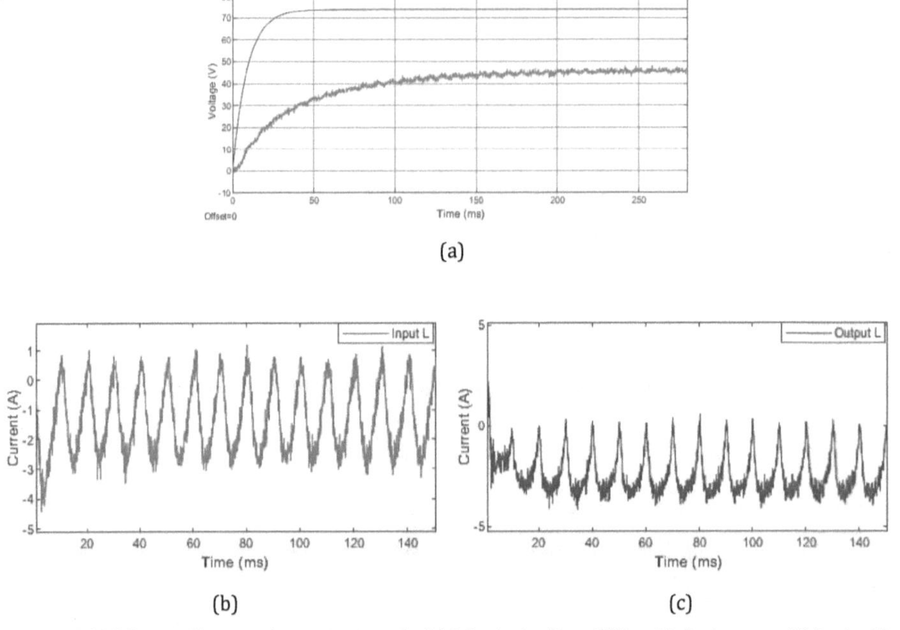


(a)

(b)

(c)

**Figure 6-7 AC-DC power flow open loop output results (a) DC output voltage (b) input inductor current (c) output inductor current**

Figure 6-7 (a) shows the output voltage only reaches a steady state value of $45.15V$. The voltage also shows larger ripples of about $7V$. The settling time is also slow only settling in $200ms$.

Table 6-3 Inverter DC-AC performance comparison between the expected and actual output voltage

| Input Step references | Steady State error (V) | Settling time (ms) | % Output ripple |
|---|---|---|---|
| **Ideal Output Voltage** | 0 | 30 | 2 |
| **Actual output voltage** | 29.85 | 198.14 | 16 |

Table 6-3 shows how the output voltage doesn't meet any of the required specifications. The ripple differs by 14% and the charging voltage is off by $29V$ with a high ripple due to the line losses, the decoupling capacitor and the filtering at the output.

### 6.4.2    *Coupling capacitor voltage*

The voltage appearing across the coupling capacitor is shown in Figure 6-8.

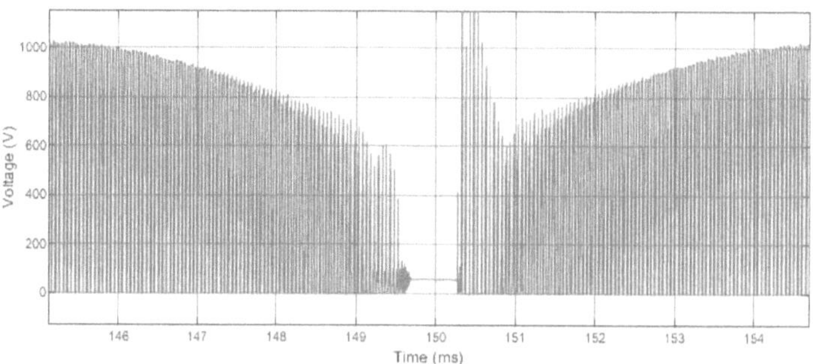

Figure 6-8 Coupling capacitor voltage

The voltage across the capacitor is discontinuous except at half cycles where it is continuous for 120us. This is due to the zero-crossing dead band discussed earlier. This continuity observed is larger than the continuity observed during simulations due to the practical dead time.

## 6.5    DC-AC power flow inverter closed loop experimental results

The inverter was then taken through a closed loop control analysis. The continuous time control loop was discretised using the backward difference equation. The sample time used was $1 \times 10^{-5}$ which is an integer multiple of the fundamental frequency used in the system. The discretised model of the control can be found in Appendix B. The output voltage and current were first tested for; then the load and line regulations and the last subsection was reserved for disturbance rejection tests.

### 6.5.1　*Output voltage*

The input voltage was kept constant at 65V as done with the open loop experimental results.

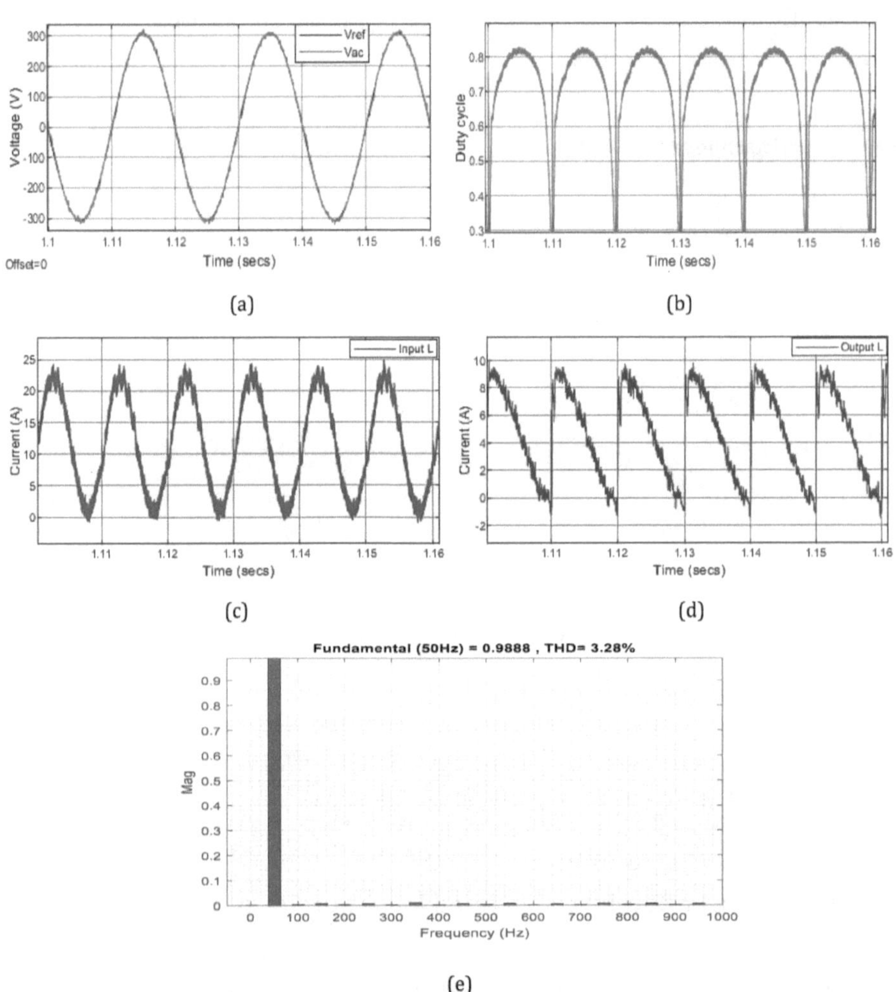

(a)

(b)

(c)

(d)

(e)

**Figure 6-9 Closed loop output performance (a) AC voltage (b) duty cycle modulating signal (c) input inductor current (d) output inductor current (e) AC output voltage FFT**

Figure 6-9 (a) shows that the output voltage now follows the reference voltage with no steady state error. The noise at the output has also significantly reduced in comparison to the open loop simulation; the effects of the decoupling capacitor were greatly negated by the control. The output voltage reaches a peak of 310.12V and an RMS value that is required at the output of the inverter. The THD of the inverter is at 3.28% from 34% in the open loop experimental results. This THD reduction value is within the required standard. The settling time for the inverter is approximately 1ms. The dead band observed at the zero crossing of the open loop experimental results has been reduced, this is because the control was designed to handle non-linearities. The duty cycle in  Figure 6-9 (b) reaches a peak of 84.36% which is required to produce the output voltage. This is slightly lower than the duty cycle peak for the simulated models. This is because a higher voltage was used since at 60V, the inverter is unable to produce the required results due to the high voltage drop across the wiring and components of the experimental set up. At duty cycles higher than this value, the output voltage produced has high harmonics and is severely distorted. At half cycles the duty cycle signal shows a jump, which is to compensate for the zero-crossing dead band observed in the open loop.

### 6.5.2    AC performance as a function of AC load power

The inverter current source was varied from an RMS amperage of $0.91A$ to $4.54A$ and the input voltage was kept constant at 65V. Table 6-4 shows the efficiency, output voltage THD and load regulation as the AC load power varies.

**Table 6-4 Inverter DC-AC performance with varying load power**

| $i_{AC,RMS}$ | $V_{AC,RMS}$ | $P_{AC}$ | $\%_{eff}$ | $\%_{THD}$ | $\%_{Load\,reg}$ |
|---|---|---|---|---|---|
| 0.916 | 220.001 | 201.521 | 64.214 | 3.245 | -2.669 |
| 1.548 | 219.985 | 340.537 | 68.141 | 3.854 | -2.669 |
| 2.058 | 219.940 | 452.637 | 78.148 | 4.125 | -2.669 |
| 2.541 | 219.870 | 558.690 | 81.234 | 4.543 | -2.670 |
| 3.015 | 218.420 | 658.536 | 79.585 | 4.698 | -2.688 |
| 3.512 | 218.102 | 765.974 | 70.246 | 4.785 | -2.692 |
| 4.069 | 217.160 | 883.624 | 62.141 | 4.859 | -2.704 |
| 4.539 | 216.129 | 971.932 | 53.115 | 5.001 | -2.742 |

Based on Table 6-4, the following graphs were plotted to show the relations between them and varying load current.

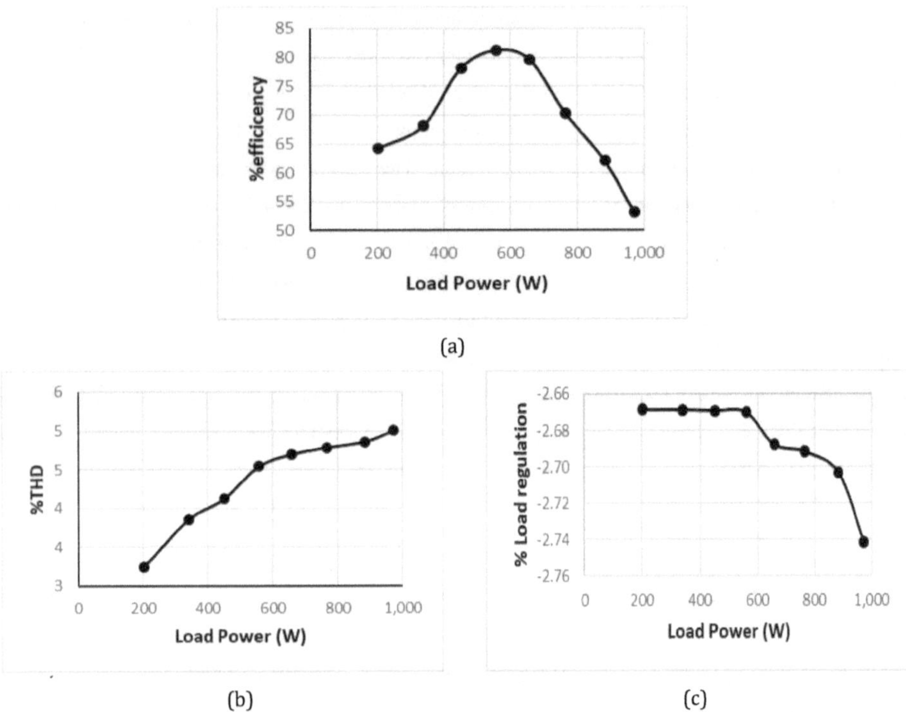



**Figure 6-10 DC-AC power flow performance with varying load current (a) efficiency (b) THD (c) load regulation**

Figure 6-10 (a) shows that the efficiency only reaching a high of 81% efficiency. The efficiency then gradually increases to reach a peak at around 2.12A – the operating point of the inverter. After this point the efficiency starts to drop off reaching a low of 74%. Figure 6-10 (b) shows the harmonic distortion increases with load power. It peaks at 5.1% which is still within the required THD standard. Figure 6-10 (c) shows that the load regulation is within the required standards.

### 6.5.3    AC performance as a function of DC line voltage

The input voltage was varied from 60 *to* 70*V* and the load current kept constant at 2.12*A RMS*, dead time at 2.7*us*. Table 6-5 shows the efficiency, output voltage ripple and the %THD during line voltage variations.

Table 6-5 Inverter DC-AC performance with varying DC input voltage

| $V_{DC}(V)$ | $V_{AC,RMS}(V)$ | $P_{AC}(W)$ | $\%_{eff}$ | $\%_{THD}$ | $\%_{Line\,reg}$ |
|---|---|---|---|---|---|
| 60.001 | 215.290 | 456.415 | 74.524 | 5.438 | 2.264 |
| 61.025 | 216.278 | 458.509 | 74.685 | 5.106 | 2.254 |
| 62.012 | 217.012 | 460.065 | 75.001 | 4.982 | 2.246 |
| 63.102 | 217.226 | 460.519 | 78.214 | 4.856 | 2.244 |
| 64.035 | 218.950 | 464.174 | 79.241 | 4.685 | 2.227 |
| 65.154 | 219.870 | 466.124 | 81.245 | 4.523 | 2.217 |
| 66.012 | 219.987 | 466.372 | 81.325 | 4.215 | 2.216 |
| 67.014 | 220.001 | 466.402 | 81.423 | 4.124 | 2.216 |
| 68.235 | 220.011 | 466.423 | 81.689 | 3.866 | 2.216 |
| 69.023 | 220.012 | 466.425 | 81.965 | 3.803 | 2.216 |
| 70.002 | 220.165 | 466.750 | 82.236 | 3.752 | 2.214 |

The following graphs show the efficiency, THD and line regulation

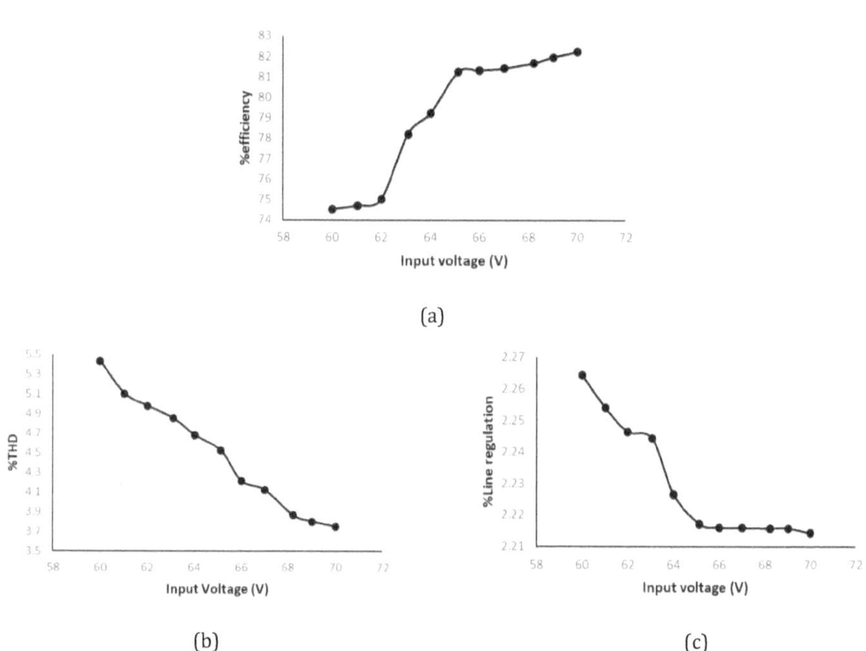

(a)

(b)         (c)

Figure 6-11 DC-AC power flow performance with varying input voltage (a) efficiency (b) THD (c) line regulation

The graph in Figure 6-11 (a) shows that as the input voltage increases, so does the efficiency. The efficiency is lowest at $60V$ input and reaches a peak of 79% around the nominal operating point of the inverter at 65V input voltage and $2.12A$ load current. This can be expected because the higher the input voltage, the lower the duty cycle needed to achieve the required RMS voltage value at the output. When the duty cycle is lower, the parasitic effects are also lower leading to less losses in gain at the output. Figure 6-11 (b) shows that the harmonic distortion improves as the input voltage rises. It is between 5.5% and 3.3% which is an accepted standard. Figure 6-11 (c) shows that the line regulation is within the required specifications.

### 6.5.4    *RMS reference set-point tracking*

The set point in the control loop was changed from 220V *RMS* to 120V *RMS* at $1.5s$ to test the tracking ability of the built inverter and digital control. The input voltage was kept constant and the load current was also unchanged.

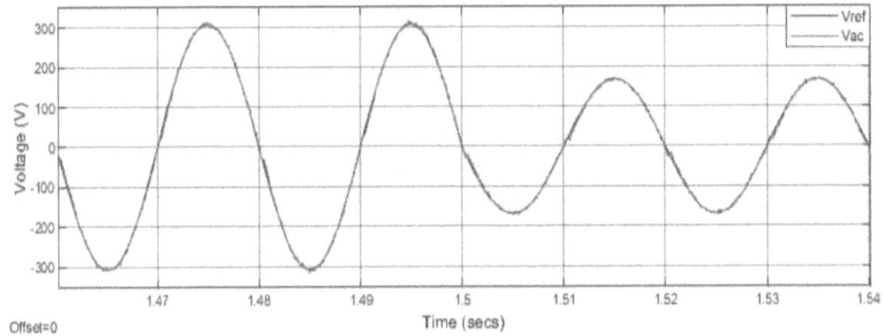


**Figure 6-12 Bidirectional inverter set point tracking**

Figure 6-12 shows that the set points were tracked with minimal steady state error. Table 6-6 shows the comparison between the two set points.

**Table 6-6 Inverter DC-AC voltage comparison between 120V *RMS* and 220V *RMS* reference voltages**

| Test Variable | Ideal Peak Value (*V*) | Peak Value (*V*) | RMS (*V*) | %THD |
|---|---|---|---|---|
| **120V *RMS* Reference** | 170 | 169.89 | 119.42 | 3.15 |
| **220V *RMS* Reference** | 311.12 | 310.17 | 216.46 | 3.28 |

The $120V$ output voltage reference had a steady state error of $0.11V$ and the $220V$ has a $0.95V$ steady state error. The THD has increased significantly in comparison to the simulated results.

### 6.5.5 Disturbance rejection tests

The inverter was put through the same disturbance tests as the simulated results. The first two were load disturbances; then input voltage disturbance and then finally, the steady state inrush current test.

### i.    Load addition disturbance rejection test

The AC load current was stepped up from $2.12A\ RMS$ to the rated current of $4.54A\ RMS$ at time $3.05s$. Figure 6-13 shows the voltage and current.

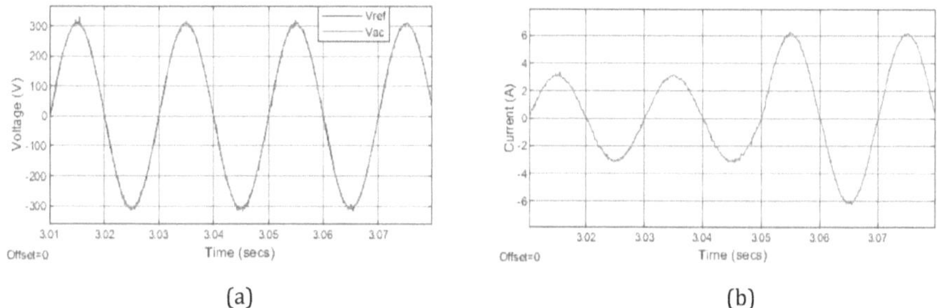

(a)                 (b)

**Figure 6-13 Load addition disturbance rejection output results (a) voltage (b) current**

Figure 6-13 (a) shows that the control successfully rejected the sudden load increase. Figure 6-13 (b) shows the load current changes appropriately and increases due to the load addition. The output voltage slightly overshoots after the load removal and shows some steady state error at rated load.

**Table 6-7 Inverter DC-AC load addition disturbance test voltage comparison between nominal and rated load**

| Test Variable | Peak Value ($V$) | RMS ($V$) | %THD |
|---|---|---|---|
| **Ideal Voltage** | 311.12 | 220 | 0 |
| **2. 12A Load** | 310.36 | 219.46 | 3.28 |
| **4. 54A Load** | 308.63 | 216.24 | 5.02 |

Table 6-7 shows that the THD rises as the load power increases. It peaks at 5.01%, when the load power is approximately $970W$. The higher load power also shows a positive steady error and increased ripples at the output.

*ii.*     ***Load removal disturbance rejection test***

The AC load current was stepped down from $2.12A\ RMS$ to $0.913A\ RMS$ at time $4.08s$. The output voltage was observed.

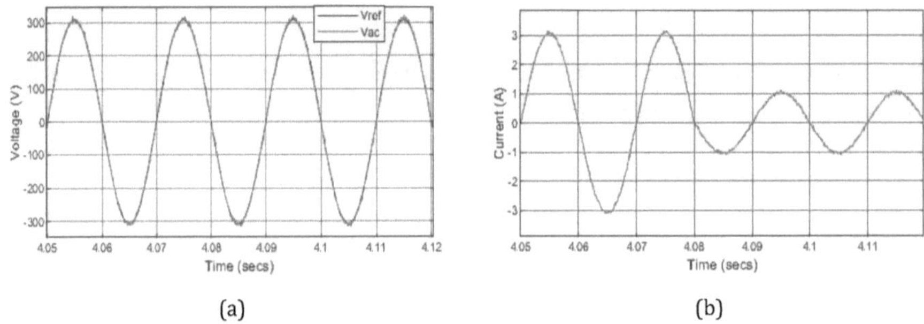

(a)                 (b)

Figure 6-14 Load removal disturbance rejection output results (a) voltage (b) current

The output voltage in Figure 6-14 (a) is shown to not be affected by the disturbance. The output current in Figure 6-14 (b) is shown to have significant ripples. This is due to the quantisation of the ADC. The filter of the output currents also plays a role in the ripples seen.

Table 6-8 Inverter DC-AC load removal disturbance test voltage comparison between nominal and minimum load

| Test Variable | Peak Value ($V$) | RMS ($V$) | %THD |
|---|---|---|---|
| **Ideal Voltage** | 311.12 | 220 | 0 |
| **2.12A Load** | 310.36 | 219.46 | 3.28 |
| **0.913A Load** | 311.24 | 220.08 | 2.75 |

Table 6-8 shows that the THD has decreased significantly to 3.27% from 4.53% and the steady state error has also gone down to just $0.2V$ due the lighter load of the system.

*iii.*     ***Input voltage disturbance test***

The input voltage was changed from $65V$ to $70V$ at time $1.5s$ to assess the ability of the control to handle sudden changes in the input voltage.

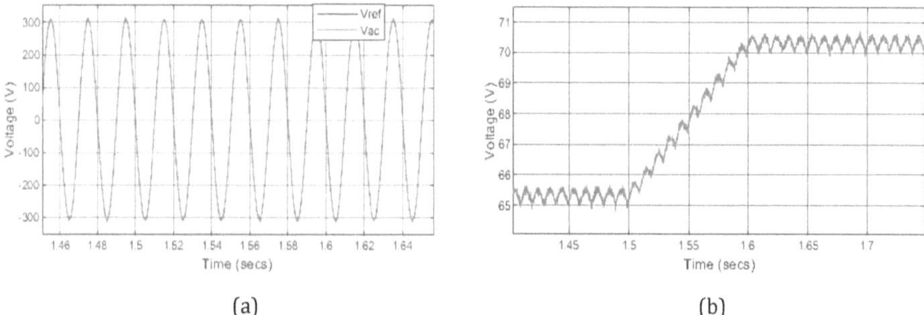

(a)                                        (b)

**Figure 6-15 Input voltage step disturbance rejection (a) AC voltage (b) Input voltage**

Figure 6-15 (a) shows that the output voltage is undisturbed. The voltage shows a reduction in ripples after the step increase. The input voltage in Figure 6-15 (b) shows the large ripples at the input due to the current drawn by the decoupling capacitor. Table 6-9 summarises the performance between the two input voltage levels.

**Table 6-9 Inverter DC-AC input voltage disturbance test voltage comparison**

| Test Variable | Peak Value ($V$) | RMS ($V$) | %THD |
|---|---|---|---|
| **Ideal Voltage** | 311.12 | 220 | 0 |
| **65$V$ Step** | 310.36 | 219.46 | 3.28 |
| **70$V$ Step** | 311.45 | 220.14 | 3.62 |

The table shows that the performance improves as the input voltage increases. The THD decreases by 25% from the lower to the higher input voltage. This is because the higher the input voltage, the lower the duty cycle required to achieve the desired AC voltage; and the lower the duty cycle, the better the performance since the parasitic elements affect the gain at higher duty cycles.

*iv.*       ***Steady state inrush currents***

The inrush currents at steady state for the inverter were tested. The rectifier load used was $50 - j0.000125\mu\Omega$ and it was introduced at $5.07s$. The inductor currents were limited between $70A$ and $-20A$.

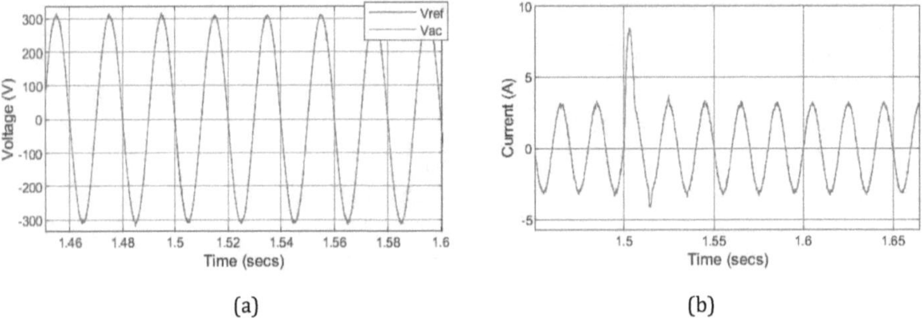

(a)                (b)

**Figure 6-16 AC Load steady state inrush currents disturbance test (a) AC voltage (b) AC current**

The load current in Figure 6-16 (b) is shown to distort and overshoot to a value of $8A$. It then settles in the next cycle. the output voltage in Figure 6-16 (a) also distorts but not as much as the current. The voltage settles within half a period, twice as fast in settling time as the current.

## 6.6    AC-DC power flow closed loop experimental results

The closed loop reverse power is presented in this section. The control loop didn't change and neither did the controller used. All the switches in the Unfolding bridge were switched off and the signals in the Ćuk converter were swapped between the input and output switches. This enabled the power to flow from the AC to the DC side. The output voltage and current were first tested. The load regulation was also checked, and the efficiency was tested. Finally, the disturbance rejection was tested.

### 6.6.1    *AC input line voltage and current*

The line current of the input AC source was shown below under a load current of $6A$ to assess the power flow and current harmonic distortion at the input in alignement with grid standards.

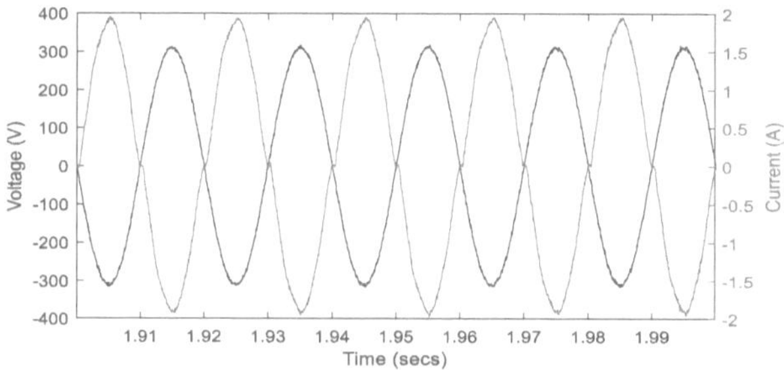

**Figure 6-17 Input line current**

Figure 6-17 shows that although the input line current still shows some distortion, it has improved from the open loop results; the THD is 6.78%. A power factor of $-9.548$ is achieved at the input. This power factor is within the desired specifications.

## 6.6.2 DC output voltage

The input AC voltage was kept constant at $220V$ RMS and a $12.33\Omega$ load resistor was used to emulate a charging current of $6A$. The output voltage was observed.

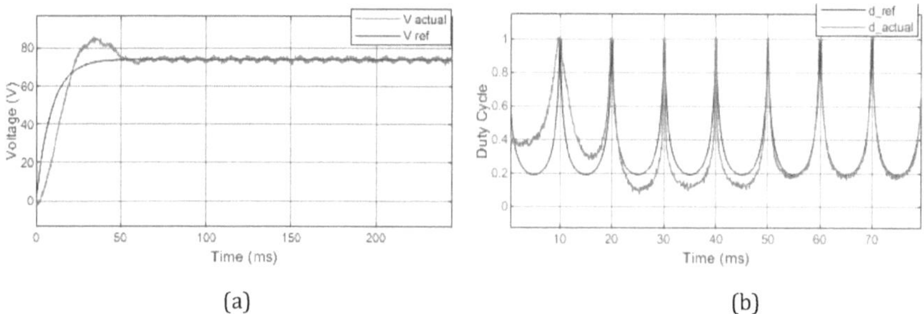

(a)          (b)

**Figure 6-18 AC-DC power flow output voltage experimental results (a) voltage (b) duty cycle modulating signal**

The output voltage in Figure 6-18 (a) is shown to have improved in its steady state value. The settling time has also improved and is now faster. The ripple is significant but less than the open loop. There is now a noticeable overshoot at the output voltage. This overshoot results as the control action tries to quicken the dynamic response of the system. The table below shows the ideal comparison. The duty cycle in Figure 6-18 (b) is shown to reach a low of 0.18. The duty cycle settles after 6 cycles as shown which is equivalent to approximately $60ms$. There are significantly ripples shown which are due to the transducer filtering. The signals also show small spikes resulting from transducer filtering.

Table 6-10 AC-DC voltage comparison between actual and ideal voltage

|  | Steady state error ($V$) | Settling time ($ms$) | % Output ripple |
|---|---|---|---|
| **Ideal Output Voltage** | 0 | 40 | 2 |
| **Actual output voltage** | 0.45 | 61.06 | 5.12 |

Table 6-10 that the steady state error reached by the voltage is $1.45V$ which is significantly less than th open loop system. The settling has improved to $61.06ms$ which is $20ms$ slower than the required. Any faster than this and the system would overshoot significantly. The ripple at the output has also dropped to 5.12% which is still higher than the expected but lower than the open results. The overshoot is 13.83% of the steady state value and is within the desired specifications.

### 6.6.3 *AC-DC performance as a function of DC load power*

The load resistor was replaced with an electronic load. With the input line AC voltage kept constant at $220V$ RMS, the load current was varied from the minimum charging current of $3A$, to the maximum charging current of $12A$. This test was done to assess the performance of the inverter at different charging currents. Table 6-11 shows the performance characteristics of the inverter as the load current varies. For the regulation column, equation (3.58) was used.

Table 6-11 AC-DC power flow performance with load varying load power

| $i_{DC}(A)$ | $i_{AC}(A)$ | $V_{DC}(V)$ | $P_{DC}(W)$ | $P_{AC}(W)$ | $\%_{eff}$ | $\%_{ripple}$ | *Settling time (ms)* | $PF$ | $\%_{THD}$ | $\%_{reg}$ |
|---|---|---|---|---|---|---|---|---|---|---|
| 3.00 | 1.44 | 73.90 | 221.76 | 316.25 | 70.12 | 5.13 | 61.06 | 0.95 | 3.92 | -2.38 |
| 4.03 | 1.90 | 73.90 | 297.45 | 417.43 | 71.26 | 5.29 | 61.51 | 0.95 | 4.10 | -2.38 |
| 5.00 | 2.26 | 73.95 | 369.91 | 498.15 | 74.26 | 5.31 | 61.65 | 0.95 | 4.52 | -2.38 |
| 6.04 | 2.69 | 74.01 | 446.66 | 592.63 | 75.37 | 5.59 | 61.81 | 0.96 | 4.98 | -2.37 |
| 7.02 | 3.15 | 74.04 | 519.95 | 693.26 | 75.00 | 5.91 | 61.97 | 0.96 | 5.21 | -2.37 |
| 8.12 | 3.68 | 73.99 | 600.82 | 809.33 | 74.24 | 6.01 | 62 | 0.96 | 5.34 | -2.37 |
| 9.01 | 4.32 | 73.88 | 665.87 | 950.93 | 70.02 | 6.13 | 62.24 | 0.97 | 5.78 | -2.38 |
| 10.02 | 4.85 | 73.79 | 739.55 | 1067.81 | 69.26 | 6.09 | 62.07 | 0.97 | 6.01 | -2.38 |
| 11.02 | 6.12 | 73.63 | 811.57 | 1347.11 | 60.25 | 6.35 | 62.2 | 0.97 | 6.09 | -2.39 |
| 12.00 | 6.67 | 72.14 | 865.54 | 1466.66 | 59.01 | 6.40 | 62.15 | 0.97 | 6.10 | -2.44 |

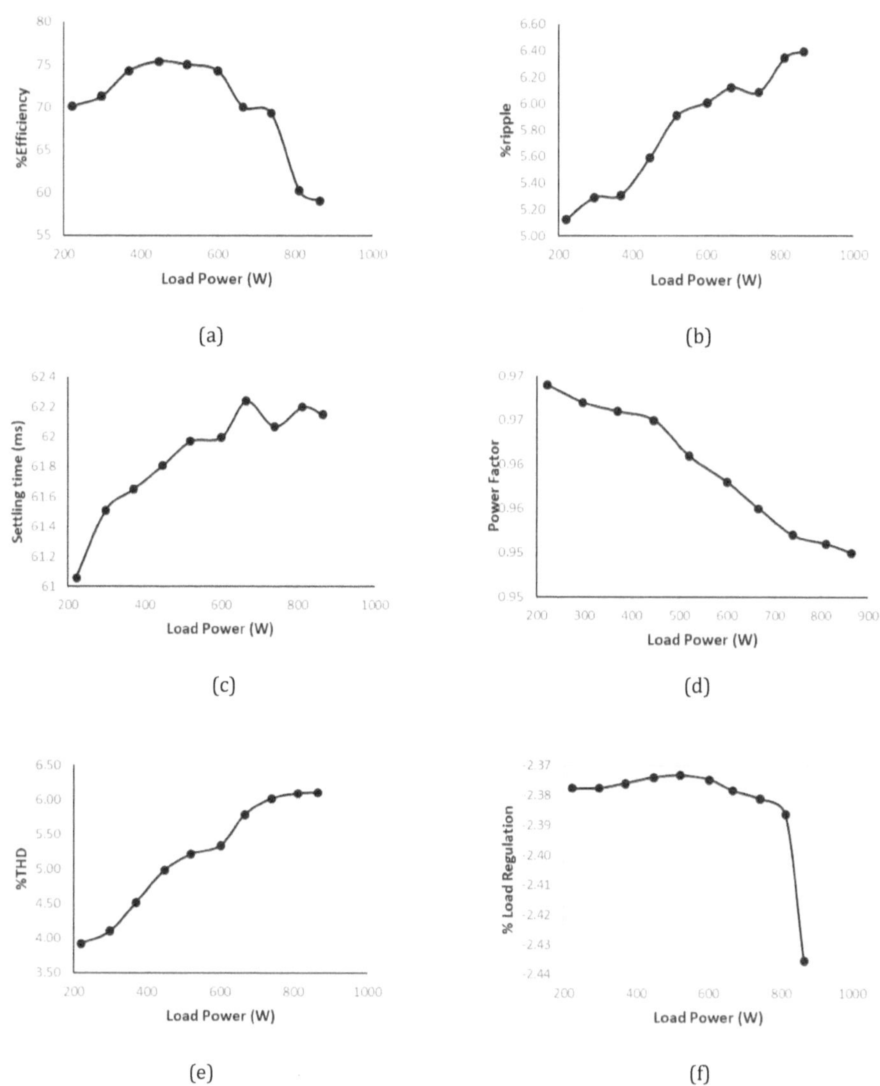

**Figure 6-19 AC-DC power flow performance as a function of load current (a) efficiency (b) ripple (c) settling time (d) power factor (e) line current THD (f) load regulation**

Figure 6-19 (a) shows that the efficiency decreases as the load power increases. It reaches a maximum of 75% at $600W$, the operating point of the inverter. Table 6-11 shows that because of the low efficiency achieved by the inverter, if a $1kW$ rated battery was used, only a maximum power of $700W$ would've been observed at the output. Figure 6-19 (b) shows the ripple increases steeply and it shown to be above the required standard of 2%. The settling time in Figure 6-19 (c) increases also but is within the desired specification. Figure 6-19 (d) shows that although the power factor decreases as the load increases, it is still a good power factor with a minimum of 0.95, within the operating range of the inverter. This decrease in power is because of the increasing line current distortion as shown in Figure 6-19 (e). The load regulation in (f) is shown to be within the quality requirements.

### 6.6.4    Set-point tracking

The input set point was changed from 74$V$ then to 64$V$ and then to 84$V$ and the output voltage was observed.

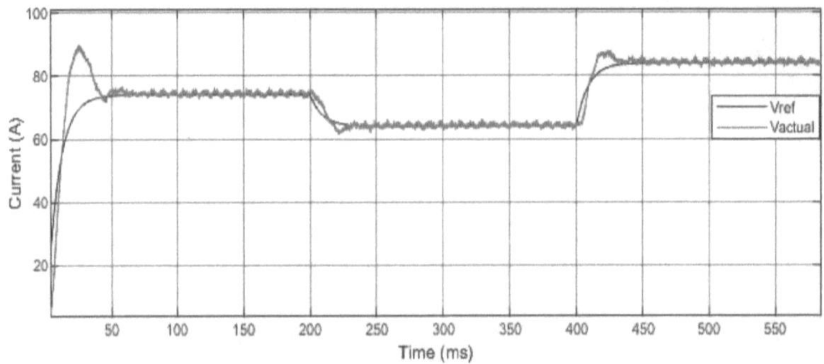


**Figure 6-20 AC-DC power flow set-point tracking**

The set points are all tracked but with some steady state error. Figure 6-20 shows that the overshoot is largest at 74$V$; the overshoot during the other two set points is minimal. Table 6-12 summarises the results.

**Table 6-12 AC-DC power flow set point output voltages comparison**

| Step references | Steady state error (V) | Settling time (ms) | % Output ripple |
|---|---|---|---|
| **64V** | 4.16 | 63.85 | 5.26 |
| **74V** | 2.45 | 61.06 | 5.12 |
| **84V** | 3.31 | 60.01 | 4.75 |

The $84V$ step performs better than the other two steps. It has the lowest ripple, the fastest settling time and lowest steady state error. The $64V$ step performs the worst having the highest ripple at the output, the slowest settling time and highest steady state error.

### 6.6.5    Disturbance rejection tests

The two disturbance tests done were load current and inrush current to compare them with the simulated results of the same tests.

#### i.    DC load current disturbance rejection assessment

The load resistor was replaced with an electronic load to emulate the battery charging stages – as done with simulated results. Figure 6-21 (b) shows that the load current was first set to $3A$ for the first $3s$; then increased to $6A$ for the next $3s$; and finally decayed back down to $3A$.

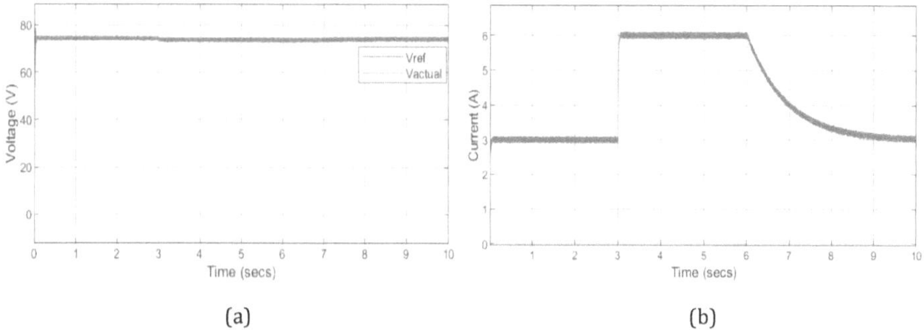

(a)           (b)

**Figure 6-21 DC load charging current disturbance rejection (a) charging voltage (b) charging current**

The charging voltage in shown in Figure 6-21 (a) remains constant during the charging duration. It shows a small dip in magnitude during the transition from pre-charge to constant current. The charging voltage level is still sufficient in charging the battery. Table 6-13 shows the performance comparison during the different charging stages.

|  | Steady state error ($V$) | Settling time ($ms$) | %Ripple |
|---|---|---|---|
| **Pre-charge** | 0.33 | 61.05 | 4.87 |
| **Constant Current** | 1.45 | 61.81 | 5.12 |
| **Constant Voltage** | ≈0.40 | - | ≈4.96 |

The charging voltage performance is shown to not meet the specifications as set out in Table 4-2. The steady state error is closest to meeting the required specification; it is less than 1% of the required value. The settling time is 33% above the required specification for all charging stages. The ripple at the output voltage is 61% above the required specification.

## ii.        *Transient Inrush currents test*

So far, the disturbance tests were done during steady state but, when transforming AC power back into DC power, it was found that there are some inrush currents at start-up if the AC voltage is applied suddenly using contacts – this leads to a large surge power at the input. This test investigates the ability of the inverter and control to compensate for these inrush currents. The filter decoupling capacitor was discharged completely before the test was performed.

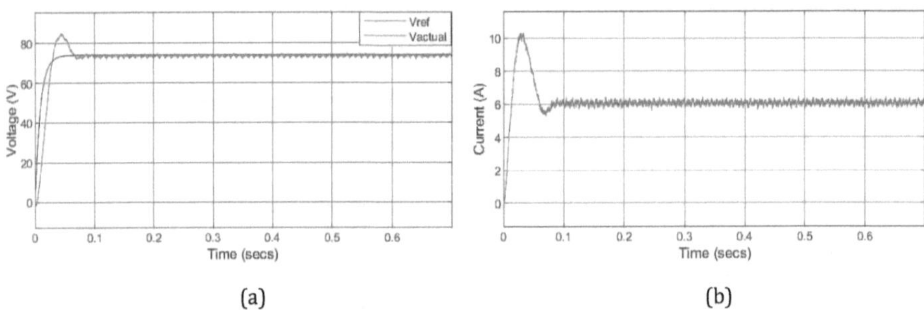

(a)        (b)

Figure 6-22 Transient Inrush currents disturbance test (a) output voltage (b) output current

The load current in Figure 6-22 (b) is shown to overshoot by 30% and oscillates for 70$ms$ before settling within 80$ms$ due to the start-up inrush currents. The output voltage in Figure 6-22 (a) shows a 10% increase in the overshoot due to the inrush currents. It settles within 81.24$ms$ and has a 67% overshoot; both are above the required specification.

# 7.  Discussions summary

Because the volume of results presented, the key areas of the simulated and the experimental results were summarized in alignment with the theory development and specifications of the inverter.


## 7.1    Output voltage performance


The inverter was designed to reach an RMS sinusoidal voltage of $220V$ – from a $60V$ to $70V$ battery DC input discharging at $1C$ – with a $2.12A$ RMS nominal AC load current during DC-AC power flow. During AC-DC power flow, it was designed to reach a steady state DC charging voltage of $74V$ from an input RMS voltage of $220V$, with a nominal $0.5C, 6A$ load charging current.

The simulation results showed that the open loop for both DC-AC and AC-DC power flows did not meet any of the required specifications as a result of the dead time and parasitic elements, as shown by Figure 5-1 and Figure 5-3. The closed loop for both power flows was able achieve the desired results with minimal no overshoot, steady state errors, ripple and THD – due to the compensation of the lead compensators of the nested loop control – as shown by Figure 5-6 and Figure 5-16.

The experimental open loop output voltages were lower than the expected and simulated results. This large drop in voltage was due to the excessive losses in the wiring and inductors – which was not a factor during simulations. During the closed loop, Figure 6-9 shows that the required 220V RMS was reached; but it required a 65V input because the dead time and parasitic elements lowered the maximum gain more than designed for. Figure 6-18 shows that the required charging voltage was reached with small low frequency ripples because of the decoupling capacitor at the input of the DC source, required to lower ripple of the input DC current. The DC output voltage also shows an overshoot that was not observed during the simulated results. This overshoot results from the fact that the control action must work harder to compensate for the slow dynamics of the transducers and filter in the experimental set up.


## 7.2    Dead time and parasitic effects


The parasitic resistances designed as $0.07\Omega$ and $0.12\Omega$ for the input and output inductors respectively, as analysed in 3.6, were supposed to ensure that the gain reached a peak value of $-5.13$ at 83.17% peak duty cycle using equation (3.117). The peak gain observed during the experimental results was $-4.61$ which was lower than the expected. This lower gain is because of the other parasitic elements such as the capacitors, switch and diodes which were neglected during the analysis of the parasitic effected gain. The dead time also lowered the gain significantly as seen from and Figure 6-18 which show the open loop only reaching a peak −2.87 gain.

## 7.3    Efficiency

The efficiency of the inverter was expected to be 94% during DC-AC power flow. The AC-DC power flow efficiency was expected to be 79.24% as analysed in section 3.7.4.

The experimental results from Table 6-5 and Figure 6-10 showed a peak efficiency of 81% during the load power variations and Table 6-5 and Figure 6-11 showed a peak efficiency of 75% for the line voltage variations, during  the DC-AC power flow. These efficiencies are 13% – 19% lower than the expected efficiency. Efficiencies greater than 90% have been achieved experimentally for both two-stage and single-stage inverters in past work. This large efficiency difference can be attributed to the fact that the inverter was built on a Vero board with wires instead of PCB which eliminates the line losses. The inductors were also manually built using wires that weren't thick enough to considerably lower the resistance; this is because manually winding a thick wire around a square ferrite core is difficult and once done, there are large gaps between the core and windings, this leads to a large leakage flux as most of the magnetic field from the wires isn't confined with the ferrite core. This leakage flux lowers the inductance value – which increases the number of turns required to reach the desired inductance value and hence, the DC resistance of the inductor which is directly proportional to the losses of the inductor. The area of the core is also larger than the inductors used from past work. Table 6-11 showed a peak efficiency of 74.9% for AC-DC power flow during load regulation. Even though this efficiency is lower than the calculated value, this efficiency is expected due to the diode H-bridge. The switch losses as analysed in section 6.1.3 had a major effect on the low efficiency of the inverter having been $27W$ for a $501W$ load. The decoupling capacitor contributed to the high-power losses as it caused large current ripples at the supply which increases the losses in power at the inductors due to the skin effect. This effect raises the inductor resistance when the large inductor current ripples flow through the outer areas of the inductor wiring.

## 7.4    Dual-Mode operation

The bidirectional inverter was designed to operate in CICM during DC-AC power flow and in DCVM during reverse AC-DC power flow from the designed passive components in section 4.3.

The simulations results showed that the during DC-AC power flow both input and output inductor currents were continuous in both open and closed loop designs for CICM. Figure 5-4 showed that the coupling capacitor voltage was discontinuous as design for to achieve DCVM. There was a slight continuity around the zero crossing for DCVM shown by the results.

The experimental results in Figure 6-8 showed that the inverter still operated in CICM mode during DC-AC power flow and operated in DCVM during reverse power flow respectively. The zero-crossing continuity was shown to have increased significantly in the experimental results. This large increase is attributed to dead time effects being more prominent during the experimental phase. The dead time increases the time that the output voltage is at zero, and hence increases the zero-crossing continuity observed.

## 7.5    Reverse power input power factor

The inverter was designed to operate in DCVM during AC-DC power flow to improve the input power factor by reducing the distortion in the input line current during AC-DC power flow; since from the theory development, this PF is inherently low for AC-DC converters operating in discontinuous mode.

The simulation results showed that the line current distortion was low, and the input power factor was almost unity. It showed a decline as the inverter was loaded. The power factor started at $-0.9998$ and dropped to $-0.9987$ during load regulation as shown by Figure 5-5. These values were within what was required from the design.

The experimental results showed the power factor was $-0.98$ and dropping to $-0.95$ during load regulation as the load power was increasing as illustrated by. This power factor decreases because the line current distortion increases as the load increases, which then lowers the power factor. This was lower than the simulated power factor but was still within power factors of $0.9 - 0.99$ achieved by power factor correction converters (PFC's) in past work.

## 7.6    Nested loop control strategy

The nested loop control strategy was designed to have good disturbance rejection and provide robust compensation for varying conditions, while the lead compensators were designed to achieve fast dynamic response as specified by Table 4-2.

The simulated closed loop results in Figure 5-6 and Figure 5-16 showed that both AC and DC output voltages were met most of the required specifications. The specification that was not met was the steady state error. The AC voltage steady state error was $0.11V$ and the DC steady state voltage $0.84V$. Although this specification was not met, in both cases these were small margins. Both AC and DC disturbances were rejected successfully within the specified settling time by the control as shown by Figure 5-10-17. The variations in the load power and line voltages were compensated for because of the G.D.C design compensations as analysed in 3.9.

The experimental AC output closed loop results in Figure 6-9 showed good tracking and a steady state error of $1.1V$, a 90% increase from the simulated results. This large increase is because of the lower than expected gain achieved by the inverter. The DC steady state error in Figure 6-18 increased by 45% from the simulated results to $1.45V$. The large error at the DC output was because of the low gain and decoupling capacitor, at the DC input, which caused large low frequency ripples. The settling time for the DC output was slower than the desired specification by 33%. This slower response was because of the transducer and filter dynamics. Both AC and DC disturbance rejections were successfully rejected as shown in Figure 6-13-26. The variations in load power and line voltage were also compensated for.


## 7.7     Load and line voltage regulations


The load regulation specification for the bidirectional inverter was $\pm 10\%$ and $\pm 5\%$ for the line regulation as specified in Table 4-1.

The simulation closed loop for the DC-AC power flow in Figure 5-8 showed load and line regulations of 0.4% and 0.8% respectively. The AC-DC power flow's load and line regulations shown in Figure 5-18 were 0.1% and 0.9% respectively – an improvement from the open loop and within the specified limits for both line and load regulations. This is because the control strategy was designed to compensated for these variations. The closed loop DC-AC power flow load regulation was 4.5% (a 4.1% increase from the simulations) and the line regulation was 1% (a 0.2% increase from the simulations) because of the filtering at the output which slightly reduced the load voltage but still within the regulation limits. The reverse power was only tested for the load regulation – since the grid voltage was assumed to not fluctuate (although in practice, grid voltage levels do fluctuate) – and it was 3.5%, a 3.4% increase from the simulated results) and still within the limits. The line regulations showed much better results than load regulations because pseudo-dc-link inverters can operate wide load current values but have the disadvantage of narrower input voltage ranges – since the inverter was only swept from a low to a high with only a $10V$ range.

## 7.8    Voltage and current total harmonic distortion

The THD specification for the bidirectional inverter was $<$ 8% for both AC Load voltage and AC input line current, as specified in Table 4-1.

The open loop simulations showed a voltage harmonic distortion of 5.93% for the ideal components with no dead time and parasitic elements for forward power flow. Once the parasitic elements were introduced the harmonic distortion increased to 8.53% which is above the required standard as shown in Figure 5-1. The closed loop simulations showed an improved harmonic distortion of 1.87%. The THD was shown to increase with increasing load current as illustrated by Table 5-3 and Table 5-4 because as the load increases, the effective output voltage decreases. It therefore improved with decreasing load. Table 5-1 showed that the THD decreased with increasing input voltage. The line current had a THD of 4.8% - which was within the required value – because of the DCVM.

The open loop experimental results showed a voltage THD of 10.41% which was above the required standard; this high THD was because of the dead time and parasitic elements. The closed loop experimental results showed a THD of an improved 4.5% because the closed loop compensates for the dead time. The AC line current was distorted during the open loop tests, with a value of 34.14% which was above the expected standard. This high THD is because of the dead time and fact that without feedback control, the coupling capacitor fails to reach a constant magnitude which leads to more current being drawn from the input supply and hence increasing the distortion. The closed loop Figure 6-17 showed an improved THD in the input line current of 7.37% which met the requirements. The THD trends discussed above were validated by Table 6-7 and Table 6-8 for the load current. Table 6-9 validates the fact that as input voltage rises the THD decreases as analysed. The line current distortion at the input side of the inverter during AC-DC power flow was shown decrease with increasing load. It reaches a maximum of 8.63% which was slightly above the required standard for THD. This was largely due to the body diodes low turn on and off times in the bridge of the inverter.

# 8.  Conclusions

Based on the discussions in the previous section, the following conclusions were drawn.

## 8.1    Satisfactory output performance

Based on the results and discussions, it can be concluded that the output voltage of the bidirectional inverter was satisfactory. Even though the experimental results didn't perform as well as the simulated results – with a low voltage and required an input voltage increase of five volts, they were still adequate, as they were within the voltage regulation range and met the required specifications with minimal steady state errors and ripples at the output.

## 8.2    Significant dead time and parasitic effects

Both simulated and experimental results showed that the dead time and parasitic effects were underestimated in the mathematical analysis because of the assumptions that were made. The results showed that these effects were significant and caused large ripples at the output. Although, the output performance was satisfactory, these effects reduced the gain by larger factor margins and caused significant dead bands at the zero crossings.

## 8.3    Poor experimental efficiency

Although, higher efficiencies have been achieved from this type of inverter, based on results, the efficiency was poor for the bidirectional inverter. The line losses from wires that were used were higher than expected. The switch losses were also higher than designed for and since the inverter had six switches, this impacted the efficiency negatively.

## 8.4    Satisfactory dual-mode performance

The inverter was able to achieve continuous inductor current mode during DC-AC power flow and was also able achieve discontinuous capacitor voltage mode during AC-DC power flow. Although the DCVM briefly showed some continuities at half cycles during zero-crossings, which were more apparent in the experimental results than the simulated results, they did not affect the inverter's power factor performance significantly as it still met its requirements.

## 8.5    Satisfactory input power factor

Based on the results, it was concluded that the power factor was satisfactory – because of the DCVM mode during AC-DC power flow as already discussed. Even though the power factor dropped lower in the experimental results, it was still within an accepted standard of range.

## 8.6    Satisfactory nested loop control design performance

A new nested loop control strategy was presented for Ćuk topology. The control strategy performed well in varying conditions and disturbance rejection tests as designed for. Based on the closed loop performance in comparison to the open loop performance and the design specifications required Although specifications for the zero-steady state error and output voltage ripple were not met because of the decoupling capacitor current, the errors were small enough that the output voltage was still well within its regulation specification.

## 8.7    Adequate line and load regulations

Even thou, the results showed that the open loop results had very poor load and line regulations, the closed loop however, showed good load and line regulations for both power flows. These regulations were within the required standards. The line regulations were better than the load regulation because the input variation was less than the load variation.

## 8.8    Satisfactory harmonic distortion

Based on the results and discussions, the THD for both line current and load voltage was satisfactory. The voltage THD was better than the line current THD since the line current THD exceeded 8% by a small margin during the experimental results.

# 9. Recommendations

From the conclusions drawn, the following recommendations were made regarding future work.

## 9.1 Use higher voltage stack

A 60V input does not reach the required peak voltage for the DC to AC inverter. A voltage of at least 65V reaches the required specifications. If batteries are used an input source, a higher voltage stack will be required to meet the specifications.

## 9.2 Include capacitive equivalent series resistances in the gain design

Consider the capacitor ESR when designing the mathematical parasitic gain ratio to get a better understanding of the expected losses and gain at the output. Even though the inductor ESR's contribute mostly to the power losses, the experimental results showed that these alone were not enough to account for the large losses seen at the output.

## 9.3 Build on a PCB and use thicker, shorter wires to improve efficiency

The circuit must be built on a PCB instead of a Veroboard. The PCB is smaller and has lower losses associated with it, therefore, the resistance decreases and allows the circuit to draw sufficient current from the supply. If the input current increases, the output current will also increase. The problem with implementing a PCB is that if a mistake is made in the connections it becomes difficult to rectify that mistake since the components are soldered onto the board. If some components must be de-soldered, it can damage the board itself of the component which means extra costs in purchasing. The other problem is that the inductors are larger when compared to other compared components so attaching them to the board becomes harder. Thick wires must be used to connect the circuit because they have less resistance than thin wires. This will also aid the increase in input current and hence the output current. The problem with using thick wires is that it is hard to implement them onto a PCB which conflicts with the recommendation above – hence, a trade off will have to be made by the designer.

## 9.4   Implement a Lead-PI compensator

Design and implement a Lead-PI compensator instead of a lead compensator. This will ensure that the closed loop steady state error is as close to zero as possible, as required, while maintaining a high bandwidth.

# 10. References


[1] E. Völkers, "Solar Power in your home," *MEDIA STATEMENT,* 2018.

[2] C. weis, "Considerations for Off-Grid PV Systems," 5 02 2013. [Online]. Available: https://www.homepower.com/articles/solar-electricity/design-installation/considerations-grid-pv-systems.

[3] Altenergy, "How do I read the solar panel specifications?," [Online]. Available: https://www.altestore.com/blog/2016/04/how-do-i-read-specifications-of-my-solar-panel/#.XGvWRegzZ9M. [Accessed 9 01 2018].

[4] A. Oros, "POWER QUALITY PERFORMANCE OF THE TRANSMISSION NETWORK IN THE REPUBLIC OF MACEDONIA," Electricity Coordinating Center, EKC , 2013.

[5] U. Power, "Alibaba," Alibaba, [Online]. Available: https://www.alibaba.com/product-detail/UPP-brand-Popular-PVC-case-70v_60380736982.html. [Accessed 9 5 2018].

[6] ellies, "ellies," [Online]. Available: https://www.ellies.co.za/product/1000w-pure-sine-wave-inverter-24vdc-230vac-50hz/. [Accessed 10 5 2018].

[7] E. P. S. M. ASSOCIATION, "Harmonic Current Emissions Guidelines to the standard EN 61000-3-2," EPSMA , 2010.

[8] J. M. a. E. Lorenzo, "ON THE SPECIFICATION AND TESTING OF INVERTERS FOR STAND-ALONE PV SYSTEMS," Instituto de Energía Solar. Universidad Politécnica de Madrid (IES-UPM), Madrid, 2017.

[9] M. Froese, "An overview of 6 energy storage methods," Maxwell Technologies, 2018.

[10] H. Energy, "Introduction to Batteries," 04 Jan 2010. [Online]. Available: http://www.ht.energy.lth.se/fileadmin/ht/Kurser/MVKF25/Batteries_Introduction.pdf.

[11] *. H. P. 1. Z. N. 1. Z. M. 1. a. C. S. 2. Lijun Zhang 1, "Comparative Research on RC Equivalent Circuit Models for Lithium-Ion Batteries of Electric Vehicles," *Open Access,* vol. 7, no. 10, p. 1002, 2017.

[12] R. a. Schwarz, "All about circuits," Rohde and Schwarz, 03 jan 2011. [Online]. Available: https://www.allaboutcircuits.com/textbook/direct-current/chpt-11/battery-ratings/. [Accessed 03 may 2018].

[13] Z. battery, "Memory Effect - What it is and what you can do about it," Z battery, 09 may 2006. [Online]. Available: http://www.zbattery.com/Battery-Memory-Effect. [Accessed 03 may 2018].

[14] C. Simpson, "BATTERY CHARGING," Texas instruments, Texas, 2005.

[15] W. Hadden, "Choose a Charger IC for Single-Cell Li-Ion Battery Applications," Texas Instruments, Texas, 2011.

[16] EPEC, "Basic Battery Charging Methods," EPEC, [Online]. Available: http://www.epectec.com/batteries/charging/. [Accessed 09 may 2018].

[17] B. a. E. Technologies, "Battery Performance Characteristics," Battery and Energy Technologies, [Online]. Available: https://www.mpoweruk.com/performance.htm. [Accessed 02 may 2018].

[18] rechargebatteries.org, "Nickel cadmium cell (NiCd)," rechargebatteries.org, 20 june 2010. [Online]. Available: https://www.rechargebatteries.org/knowledge-base/batteries/nickel-cadmium-cell-nicd/. [Accessed 05 may 2018].

[19] N. B. A. a. Disadvantages. [Online]. Available: https://www.doityourself.com/stry/nicd-battery-advantages-and-disadvantages.

[20] B. university, "Nickel-based Batteries," battery university, 01 jan 2006. [Online]. Available: http://batteryuniversity.com/learn/article/nickel_based_batteries. [Accessed 02 april 2018].

[21] b. university, "BU-204: How do Lithium Batteries Work?," battery university, 01 jan 2006. [Online]. Available: http://batteryuniversity.com/learn/article/lithium_based_batteries. [Accessed 02 may 2018].

[22] N. Mohan, "dc-dc switch mode converters," in *Power Electronics, 2nd edition*, Canada, 1995, p. 184.

[23] M. SHAYESTEGAN, "Overview of grid-connected two-stage transformer-less inverter design," *Journal of Modern Power Systems and Clean Energy,* vol. 6, no. 4, pp. 642-655, 2018.

[24] L. C. Y. X. J. Bordonau, "Topologies of Single-Phase Inverters for SmallDistributed Power Generators," *IEEE TRANSACTIONS ON POWER ELECTRONICs,* p. 13, 2004.

[25] T. K. e. al, "A New High-efficiency Single-Phase Transformerless PV Inverter toplogy," *IEEE Tran. Ind. Electron.,* vol. 58, no. 1, pp. 184-191, 2011.

[26] Y. B. L. Y. W. Jiuchun Jiang, "Topology of a Bidirectional Converter for Energy Interaction between Electric Vehicles and the Grid," *Annual Power Electronics, Drives Systems and Technologies Conference.*

[27] M. C. a. V. G. A. Minsoo Jang, "A Compact Single-Phase Bidirectional Buck-Boost-," *IEEE,* 2016.

[28] T. A. L. Jun Kikuchi, "Three-Phase PWM Boost–Buck Rectifiers With Power-Regenerating Capability," *IEEE TRANSACTIONS ON INDUSTRY APPLICATIONS,* vol. 35, no. 5, 2002.

[29] shodganga, "Survey on Sliding Mode Control".

[30] M. O. I. H. M. E. J. A. D. Abreu-Garcia, "SLIDING MODE CONTROL OF THE CUK CONVERTER," *PESC Record. 27th Annual IEEE Power Electronics Specialists Conference,* 1996.

[31] S. S. Jeremy Knight, "An Improved Reliability Cuk Based Solar Inverter With Sliding Mode Control," *IEEE TRANSACTIONS ON POWER ELECTRONICS,* vol. 21, no. 6, 2006.

[32] S. Keeping, "Voltage- and Current-Mode Control for PWM Signal Generation in DC-to-DC Switching Regulators," *Electronic Products,* 01 10 2014.

[33] L. Dixon, "Average Current Mode Control of Switching Power Supplies," *unitrode application note,* vol. 140.

[34] C. C. A. G. Mihail Octavian Cernaianu, "Ćuk converter employing indirect current control loop for TEG energy harvesting devices," *2012 IEEE 18th International Symposium for Design and Technology in Electronic Packaging (SIITME),* 2012.

[35] A. A. A. a. J. Nazarzadeh*, "Improvement Behavior and Chaos Control of Cuk Converter in Current Mode Controlled," *2008 IEEE International Conference on Industrial Technology,* pp. 1-6, 2008.

[36] B.-T. Lin and Y.-S. Lee, "Power-factor correction using Cuk converters in discontinuous-capacitor-voltage mode operation," *IEEE Transactions on Industrial Electronics,* vol. 44, no. 5, pp. 648 - 653, 1997.

[37] E. team, "Analysis of Four DC-DC Converters in Equilibrium," All about circuits, 06 june 2015. [Online]. Available: https://www.allaboutcircuits.com/technical-articles/analysis-of-four-dc-dc-converters-in-equilibrium/. [Accessed 01 june 2018].

[38] B.-T. L. a. Y.-S. Lee, "Power-Factor Correction Using Cuk Converters in Discontinuous-Capacitor-Voltage Mode Operation," *IEEE TRANSACTIONS ON INDUSTRIAL ELECTRONICS,,* vol. 44, no. 5, pp. 648-653, 1997.

[39] B.-T. L. a. Y.-S. Lee, "Power-Factor Correction Using Cuk Converters in Discontinuous-Capacitor-Voltage Mode Operation," *IEEE TRANSACTIONS ON INDUSTRIAL ELECTRONICS,* vol. 44, no. 5, 1997.

[40] H. IO, "PWM control and Dead Time Insertion," 12 09 2014. [Online]. Available: https://hackaday.io/project/3176-gator-quad/log/11741-pwm-control-and-dead-time-insertion. [Accessed 06 06 2018].

[41] B. K. K. a. M. A. Narain, "Controller Design for Cuk Converter Using," *International Conference on Power, Control and Embedded Systems,* 2012.

[42] H. H. e. al, "Decoupling techniques for micro-inverters in PV systems - a review".

[43] J. D. Kramer and C. Jacky, "How to write biblos," vol. 1, no. 1, 2006.

[44] T. electronics, "technics electrochemistry," technics, 01 feb 2010. [Online]. Available: http://www.homofaciens.de/technics-electrochemistry-rechargeable-cells_en.htm. [Accessed 01 may 2018].

[45] shodganga, "Batteries general information," bitstream, 01 june 2005. [Online]. Available: http://shodhganga.inflibnet.ac.in/bitstream/10603/1303/7/07_chapter%201.pdf. [Accessed 01 may 2018].

[46] Cadex, "BU-402: What Is C-rate?," Battery university, 9 may 2009. [Online]. Available: http://batteryuniversity.com/learn/article/what_is_the_c_rate. [Accessed 06 may 2018].

[47] M. mania, "Types of battery," Mechanical, 02 April 2013. [Online]. Available: http://mechanicalmania.blogspot.com/2011/07/types-of-battery.html. [Accessed 02 may 2018].

[48] D. linden, "Rechargeable zinc/alkaline/manganese dioxide batteries," in *Handbook of Batteries Third Edition*, Mcgraw Hill, 2002, pp. 36-100.

[49] S. Krishnamurthi, "A Tale of Two Batteries - Primary and Secondary Alkaline Battery," *Frost & Sullivan Market Insight,* vol. 8, no. 10, p. 66, 2003.

[50] J. Donovan, "Battery Management System for Charge CCCV, Inc.," *Zhang's Research Group,* 2003.

[51] K. W. E. C. a. S. L. H. ZHANGHAI SHI†, "BOUNDARY CONDITION ANALYSIS FOR CUK, SEPIC," *Journal of Circuits, Systems, and Computers,* vol. 22, no. 01, pp. 1250067-1 1250067-10, 2011.

[52] E. M. K. DAS, "Sine Wave Generation with PIC micro-controller," M's Lab, 17 jab 2015. [Online]. Available: https://mlabsbd.wordpress.com/tag/pure-sine-wave-inverter/. [Accessed 02 may 2018].

[53] V. michal, "Three-Level PWM Floating H-Bridge Sinewave Power Inverter for High-Voltage and High-Efficiency Applications," *IEEE transactions on Power,* pp. 1-9, 2015.

[54] S. V. R.Ravikumar3, "Implementation of Direct AC-DC Boost Converter For Low Voltage Energy Harvesting," *International Journal of Engineering Science Invention,* vol. 2, no. 3, pp. 08-16, 2013.

[55] S. A.-R. F. S. K. Siu, "PFC boost converter design guide," 01 may 2006. [Online]. Available: https://www.infineon.com/dgdl/Infineon-ApplicationNote_PFCCCMBoostConverterDesignGuide-AN-v02_00-EN.pdf?fileId=5546d4624a56eed8014a62c75a923b05. [Accessed 09 april 2018].

[56] E. team, "Analysis of 4 dc-dc converters in equilibrium," All about circuits, 06 june 2015. [Online]. Available: https://www.allaboutcircuits.com/technical-articles/analysis-of-four-dc-dc-converters-in-equilibrium/. [Accessed 23 december 2018].

[57] B. Poorali, E. Adib and H. Farzanehfard, "Soft-switching DC–DC Cuk converter operating in discontinuous-capacitor-voltage mode," *IEEE IET Power Electronics,* vol. 13, no. 10, pp. 1679 - 1686, 2017.

[58] A. U. E. G. !. !. a. L. M. P. Sanchis, "Design and experimental operation of a control," *IEEE,* 2005.

[59] A. Namboodiri, "Unipolar and Bipolar PWM Inverter," *International Journal for Innovative Research in Science & Technology,* vol. 1, no. 7, pp. 237-244, 2014.

[60] E. Team, "Analysis of Four DC-DC Converters in Equilibrium," 06 June 2015. [Online]. Available: https://www.allaboutcircuits.com/technical-articles/analysis-of-four-dc-dc-converters-in-equilibrium/. [Accessed 1 June 2018].

[61] J. Moraka, "Dead time effects on Boost inverter," IEEE, Cape town, 2014.

[62] T. Mandi´c, "DC/DC Converter Dead-Time Variation Analysis Far-Field Radiation Estimation," *EMC Compo*, 2015.

[63] 0. A. S. a. L. M. P. P. Sanchis Gdrpide, "A New Control Strategy for the Boost DC-AC Inverter," *IEEE*, pp. 974-984, 3014.

[64] E. Co, "Lithium-ion Battery," 2010. [Online]. Available: https://www.ineltro.ch/media/downloads/SAAItem/45/45958/36e3e7f3-2049-4adb-a2a7-79c654d92915.pdf. [Accessed 01 may 2018].

[65] M. M.-U.-T. Chowdhury, "A NOVEL SINGLE-STAGE INVERTER TOPOLOGY," *Northeastern University*.

[66] N. A. I. S. Suat Ozdemir, "Single stage three-level MPPT inverter for solar supplied systems," *International Symposium on Power Electronics Power Electronics, Electrical Drives, Automation and Motion*, 2012.

[67] J. Chen, "A Bidirectional Single-Stage DC/AC Converter for Grid Connected Energy Storage Systems," *Journal of Power Electronics*, pp. 1026-1034, 2015 .

[68] K. Cheng, "Boundary condition analysis for Cuk, Sepic and Zeta," *Journal of Circuits, Systems and Computers*, 2013.

[69] E. H. Ismail, "A High-Quality Rectifier Based on Sheppard–TaylorConverter Operating in DiscontinuousCapacitor Voltage Mode," *IEEE TRANSACTIONS ON INDUSTRIAL ELECTRONICS,*, vol. 55, no. 1, p. 38, 2008.

[70] Shodganga, "POWER FACTOR CORRECTION CONVERTERS," Shodganda, [Online]. Available: http://shodhganga.inflibnet.ac.in/bitstream/10603/42171/7/07_chapter2.pdf.

[71] G. Knier, "How do Photovoltaics Work?," 06 08 2008. [Online]. Available: https://science.nasa.gov/science-news/science-at-nasa/2002/solarcells. [Accessed 8 05 2018].

[72] N. A. I. A. C. Z. T. Nahidul Hoque Samrat1, "Technical Study of a StandalonePhotovoltaic–Wind Energy Based HybridPower Supply Systems for IslandElectrification in Malaysia," *Centre for Product Design and Manufacturing (CPDM*, pp. 2-65, 2015.

[73] S. p. sysytems, "medium residential off grid solar system," Specialised power systems, 6 4 2015. [Online]. Available: https://specializedsolarsystems.co.za/product-catalogue/ac-solar-power-solutions/complete-off-grid-solar-system-solutions/5kva-18-24kwh-per-day-off-grid-solar-system/. [Accessed 8 6 2018].

[74] C. A. authority, "Techinical guidance material electrical load analysis," Civil Aviation authority, 2013.

[75] S. Keeping, "A Review of Zero-Voltage Switching and its Importance to Voltage Regulation," Didikey, 2014.

[76] G. T. D. S. Yasmeena*, "A Review of Technical Issues for Grid Connected Renewable Energy Sources," *International Journal of Energy and Power Engineering*, vol. 4, no. 5-1, pp. 22-32, 2015.

[77] A. K. a. V. John, "HF Transformer Based Grid-Connected Inverter Topology for Photovoltaic Systems," *IETE*, 2015.

[78] J. K. a. N. E.Koutroulis, "A bidirectional, sinusoidal, high-frequency inverter design," *IEEE Elctric power applications*, 2001.

[79] M. B. A. G. a. A. M. Tarak Salmi#1, "Analysis of Harmonics and Common-Mode Voltage in Transformerless AC Modules Integrated in PV System," *Renewable Energies and Electric Vehicles, National Engineering*, pp. 589-596, 2012.

[80] P. t. systems, "Lithium-Ion State of Charge (SoC) measurement," Power tech, [Online]. Available: https://www.powertechsystems.eu/home/tech-corner/lithium-ion-state-of-charge-soc-measurement/. [Accessed 03 06 2018].

[81] F. Ahmad, A. Rasool, E. Ozsoy, A. Sabanovic and M. Elitas, "Design of a robust cascaded controller for Cuk," *IEEE International Power Electronics and Motion Control Conference, Varna,*, pp. 80-85, 2016.

[82] S. Tan, Y. Lai and K. Chi, "General design issues of sliding-mode controllers in DC-DC converters," *IEEE Trans. Power Electron,* vol. 55, pp. 1160-1174, 2008.

[83] R. Venkataramanan and Sabanovic, "C′ uk, S. Sliding mode control of DC–DC converters," *IEEE 51st Annual Conference on Decision and Control, Maui, HI, USA,* 2012.

[84] H. W. G. a. J. H. J. Y, "Variable Structure Control: A Survey," *IEEE,* vol. 40, no. 1, 1993.

[85] F. Ahmad, A. Rasool, E. Ozsoy, A. Sabanovic and M. Elitas, "A robust cascaded controller for DC-DC boost and C′ uk converters. World J. Eng.," *2017,* vol. 14.

[86] F. B. O. G. David Meneses, "Review and Comparison of Step-Up Transformerless Topologies for Photovoltaic AC-Module Application," *IEEE TRANSACTIONS ON POWER ELECTRONICS,* vol. 28, no. 6, pp. 2649-2658, 2013.

[87] N. S. Y. M. A. Q. H. Haifei Deng Xiaoming Duan, "Monolithically Integrated Boost Converter," *IEEE TRANSACTIONS ON POWER ELECTRONICS,* vol. 20, no. 3, 2005.

[88] D. G. B. Z. I. Daho*, "Stability Analysis and Bifurcation Control of Hysteresis Current Controlled Cuk Converter Using Filippov's Method," *2008 4th IET Conference on Power Electronics, Machines and Drives,* pp. 381-385, 2008.

www.ingramcontent.com/pod-product-compliance
Lightning Source LLC
LaVergne TN
LVHW042110190726
843493LV00006B/1436